T0133328

Fabian Russell Pfitz

Trajectory tracking, path following, and learning in model predictive control

Logos Verlag Berlin

λογος

Bibliografische Information der Deutschen Nationalbibliothek

Die Deutsche Nationalbibliothek verzeichnet diese Publikation in der
Deutschen Nationalbibliografie; detaillierte bibliografische Daten sind
im Internet über http://dnb.d-nb.de abrufbar.

D 93

ISBN 978-3-8325-5705-8

Logos Verlag Berlin GmbH
Georg-Knorr-Str. 4, Geb. 10,
D-12681 Berlin
Germany

Tel.:   +49 (0)30 / 42 85 10 90
Fax: +49 (0)30 / 42 85 10 92
http://www.logos-verlag.de

Table of Contents

Acknowledgement

My thesis was carried out as cooperation project between the Institute for Systems Theory and Automatic Control (IST) at the University Stuttgart and the company Porsche Engineering Services GmbH (PES) in Bietigheim-Bissingen. First of all, I like to say thank you that I had the rare opportunity to do both institutional work and participating in series development. I can recommend this constellation to all the persons out there who are willing to learn much, work hard and above all have fun.

Most thanks go to my professor Christian Ebenbauer who accompanied me the whole way of my phd thesis. Thank you for discussing with me on an excellent professional level. He enlarged my horizon both on a personal and professional level, I will always keep our time at the IST in mind, thank you, I hope we will always stay in contact! Many thanks to the members at the IST. Due to the cooperation project, I have to admit I was rarely at the institute. Nevertheless, I really liked the cohesion and the work spirit of the whole team. Above all, I really enjoyed the Christmas party and our seminars with presentations about the current reseach areas. Special thanks go here to Stefan Wildhagen, Anne Romer and Julian Berberich.

Also many thanks to my supervisor Dr. Martin Braun at Porsche Engineering Service GmbH. He always encouraged me to push further. I really appreciate his unique management style, but even more his unique personality. Also thanks to the colleagues Steffen Sailer and Max Schäfer who became good friends. Many thanks to my room colleague and close friend Jan Feiling. We know each other from the very beginning of our studies. The joint time at the IST even made our friendship closer, I like to say thank you for your sympathic, trustful and excellent personality.

Special thanks go to my parents, Manfred and Monika, and my brother Stefan who always supported me. Thank you that you gave me the opportunity to follow and realize my dreams. Thanks for raising me up to be curious about developing new things and the encouragement with respect to engineering and new technologies from the very beginning. Last, I like to thank my wife Lucia and my cute daughter Valentina, who gave me power and motivation whenever I needed it, you are the best thing in my life, I will always love you.

Abstract

In this thesis, we present novel model predictive control (MPC) formulations based on a convex open-loop optimal control problem to tackle the problem setup of *trajectory tracking* and *path following* as well as the control of *systems with unknown system dynamic*. In particular, we consider the framework of relaxed barrier function based MPC (rbMPC). We extend the existing stability theory to the trajectory tracking and the path following problem. We establish important system theoretic properties like closed-loop stability and exact constraint satisfaction under suitable assumptions. Moreover, we evaluate the developed MPC algorithms by comparing it with a standard controller in the area of automated driving in simulations as well as in real-world for an automated driving scenario. Further, we consider the control of completely unknown systems based on online optimization. We divide the overall problem into the design of an estimation algorithm and a control algorithm. The proposed estimation algorithm does not belong to the class of identification algorithms but can rather be seen as an asymptotically accurate *signal predictor algorithm* of the closed-loop trajectory. The control algorithm is a model-independent receding horizon control algorithm in which important system theoretic properties like convergence to the origin are guaranteed without the knowledge of the true system parameters. The estimation and control algorithm are combined together and convergence to the origin of the closed-loop system for fully unknown linear time-invariant discrete-time systems is shown.

Deutsche Kurzfassung

In der vorliegenden Arbeit werden neue, auf konvexen Optimierungsproblemen basierende modellprädiktive Regelungsverfahren (engl. model predictive control, MPC) untersucht. Ziel ist die Entwicklung von Algorithmen für das *Trajektorienfolgeproblem* und *Pfadverfolgungsproblem* sowie der Regelung von Systemen ohne Kenntnis eines Systemmodells. Dabei erweitern wir die existierende MPC Theorie, basierend auf relaxierten Barrierefunktionen, auf die Problemstellung des Trajektorien- und Pfadverfolgungsproblems. Hier weisen wir wichtige systemtheoretische Eigenschaften wie die Stabilität des geschlossenen Kreises und die exakte Einhaltung von Beschränkungen unter bestimmten Annahmen nach. Zudem vergleichen wir die entwickelten MPC Algorithmen simulativ mit einem Standardregler im Bereich des automatisierten Fahrens und werten Daten einer experimentellen Erprobung aus. Außerdem betrachten wir die optimierungsbasierte Regelung von Systemen ohne Kenntnis eines Systemmodells. Der Lösungsansatz basiert auf dem Entwurf eines Schätzalgorithmus und eines Regelalgorithmus. Dabei gehört der Schätzalgorithmus nicht zur Klasse der Systemidentifikationsalgorithmen, sondern kann als asymptotisch korrekter Signalprädiktor der Trajektorie des geschlossenen Kreises aufgefasst werden. Der Regelalgorithmus gehört zur Klasse der MPC Algorithmen, welcher Konvergenz des geschlossenen Kreis zum Ursprung ohne genaue Kenntnis der Systemparameter garantiert. Schließlich kombinieren wir den Schätz- und den Regelalgorithmus und weisen Konvergenz des geschlossenen Kreis zum Ursprung für linear zeitinvariante Systeme nach.

1
Introduction

1.1 Motivation

Autonomously driving vehicles are one of the main scientific and technological challenges of our time. Basically, there exist different automation levels, starting from no automation to full automation of the vehicle in which the surveillance of the driver is not needed anymore, see E. Yurtserver et al. (2020). From a conceptual point of view, the architecture of autonomous driving consists of three steps: *Sense, understand and plan*, and *act*. Sensing refers to the first step and deals with the transfer of environmental data into a digital map based on the available vehicle sensors, e.g., cameras, radars or lidars. Understanding and planning depicts the second step and mainly takes care about the calculation of a safe trajectory based on the digital map. Finally, the acting step considers the control of a vehicle by steering, braking and accelerating commands such that the deviation to the trajectory, calculated in the sensing step, is minimal. The topic addressed in this thesis is motivated by automated driving on test tracks. In particular, we like to focus on the development process of a vehicle. There, the vehicle has to successfully pass several tests, e.g., endurance tests, high speed tests or braking tests before it is delivered to the customer. These tests are carried out by trained drivers on either public streets or non-public proofing grounds to ensure public safety. In case of non-public proofing ground tests, the driver typically aims to drive a given driving line as reproducible as possible in a repetetive manner. However, given that the driver only remembers the most important points of the test run, he or she will make non-reproducible errors. Hence, it is difficult to locate the error source that originates from either the test driver or a defective part of the vehicle. Thus, it is desirable to tackle this problem by replacing the driver by an algorithm tailored for automated driving to ensure reproducability between the different test runs.

In this thesis, we apply MPC to certain tasks in an automated driving scenario. We will use here the term automated driving and not autonomous driving, given that we only deal with the acting step of the autonomous driving architecture. MPC, also referred to as *receding horizon control*, is a model-based control method in which a suitable control input is obtained by solving online an open-loop optimal control problem. Herein, a user-defined cost criterion along a specific prediction horizon is minimized under the explicit consideration of certain state and input constraints. Owing to the flexibility of the user-defined cost criterion, many industrial control tasks have been so far addressed using an MPC formulation such as the control of an industrial servo machine tool drive, see M. Stephens and M. Good (2013), the control of a chemical reactor, see M. Bakosova and J. Oravec (2014), or the control of an automated driving vehicle, see B. Gutjahr, L. Gröll and M. Werling (2017). For a survey paper about the challenges and opportunities of MPC in industrial applications, we refer to M. Forbes et al. (2015). A driver acts in many ways similar to MPC. Not only does MPC naturally provide a prediction horizon, it also enables us to include the vehicle dynamics as well as physical constraints such as acceleration constraints. Thus, it is a natural choice to use MPC in auomated driving scenarios. Nevertheless, standard MPC formulations have two disadvantages. On the one hand, the online solution of the open-loop optimal control problem requires high computational effort but at the same time has to satisfy hard real-time requirements. This might be problematic implementing the algorithm on low-cost ECUs (electronic control unit). On the other hand, consider a real system in which, in almost any case, the prediction model of the MPC formulation is not exactly capturing the real system behaviour. Operating at the limit of the imposed constraints, the mismatch between the prediction model and the actual system behaviour may lead to constraint violation and thus to an infeasible optimization problem and instabilities. One promising approach to overcome these issues is *relaxed barrier function* MPC (rbMPC), which enjoys desirable properties like convergence guarantees under suboptimal inputs, robustness properties, exact constraint satisfaction under certain initial conditions, and feasibility even if the constraints are violated. For an overview of this topic, we refer the reader to C. Feller and C. Ebenbauer (2016, 2017, 2020) and C. Feller (2017). However, so far, a proper control and systems theoretic investigation of an rbMPC setup tailored for the problem class of automated driving is not available in literature. Hence, we will investigate rbMPC formulations which can be used for automated driving. In particular, we focus on two main problem formulations to follow a desired driving line at a specific velocity profile. One is the so-called *trajectory tracking* problem, and the other one refers to the so-called *path following* problem. In a trajectory tracking problem setup, one tries to track a time-varying trajectory in which the timing of the trajectory is implicitly predefined by the trajectory itself. In the path following problem, one likes to track a geometric path where the timing on this geometric path is left an additional degree of freedom to the controller. Path following allows, e.g., to overcome certain performance limitations, see A. Aguiar, J. Hespanha and P. Kokotovic (2005) for nonminimum phase systems in case of unstable zero dynamics. Due to the fact that some of the applied control energy must be used for its stabilization.

The control performance of an MPC algorithm is highly depending on the used prediction model, or, in terms of the driver, directly corresponds to how well he or she knows the vehicle dynamics. The better the prediction model fits the actual system behavior, the better the transient behavior of the closed loop will be. However, models are never exact and become often uncertain where modeling of the decisive effects is hard or expensive in time and cost. This is true for a wide range of industrial applications. In the case of an automated driving vehicle, this is important for the high dynamics area, where for example, the tire slip curve is located in its nonlinear region, see for example H. Pacejka (2012) and thus the tire model becomes inaccurate. Hence, this motivates us to investigate an MPC formulation where the prediction model is learned online via input and output data. Many different approaches exist in the adaptive control and learning literature consisting of model-free approaches and model-based ones, see G. Goodwin and K. Sin (2009); G. Tao (2014); M. Benosman (2016). A quite common procedure in the control of unknown systems is to divide the overall control task into a control scheme and an estimation scheme. This strategy is also pursued in this thesis. However, in contrast to the existing literature, see P. Tabuada and L. Fraile (2020); T. Nguyen et al. (2020); V. Adetola, D. DeHaan and M. Guay (2009), we develop a fully online optimization-based solution for the estimation and control scheme with provable convergence to the origin of the closed-loop system for completely unknown linear time-invariant discrete-time systems.

1.2 Contribution and outline

In this section, we will summarize the main contributions of the thesis and we will give a brief outline of the thesis's structure. The main contributions consist of two parts. In the first part, in Chapter 2, we extend the theory of relaxed barrier functions to the *trajectory tracking* and *path following* problem. We exploit the properties of relaxed barrier functions to guarantee important system-theoretic properties like *closed-loop stability* and *exact constraint satisfaction* for a certain set of initial conditions under rather mild assumptions for linear time-invariant discrete-time systems. Further, we apply both algorithms to numerical examples to illustrate the theoretical results. In the second part, in Chapter 3, we consider the stabilization of the origin of fully unknown linear time-invariant discrete-time systems. We divide the problem into an *estimation problem* and a *control problem*. For the estimation algorithm, we design an optimization algorithm based on a proximal minimization algorithm. The estimates do not necessarily converge to the true system parameters and thus the estimator does not belong to the class of model identification algorithms but can rather be seen as an asymptotically accurate *signal predictor algorithm* of the closed-loop trajectory. For the control algorithm, we design a new model-independent receding horizon control scheme. We prove convergence to the origin for a certain class of prediction models without the knowledge of a system model. Therefore, we refer to the control algorithm as a *modeling-free* receding horizon control policy. Further, we combine both schemes into a full online optimization control scheme and show convergence to the origin of the closed loop under certain assumptions. We provide a strict convergence analysis to the origin for unknown linear time-invariant discrete-time systems. Finally, we show the potential of the proposed algorithm for nonlinear systems.

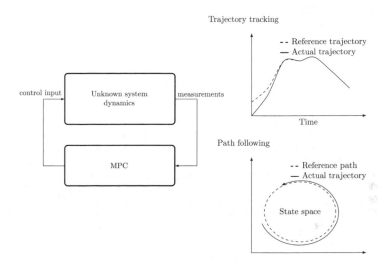

Figure 1.1. Summary of the main ingredients of the thesis. Left: MPC applied to systems with unknown system model. Right: Conceptual illustration of trajectory tracking and path following. In trajectory tracking, the control goal is to track an a priori given reference trajectory. In path following, the control goal is to follow a parametrized curve in the state space.

Figure 1.1 summarizes the two main contributions of the thesis. On the one hand, it shows the high level signal flow of an MPC scheme being able to control a system with fully unknown system dynamics. On the other hand, it highlights the difference between *trajectory tracking* and *path following*.

Chapter 4 deals with *simulation tests*, *modeling*, and the *efficient implementation* of the trajectory tracking and path following controller based on rbMPC, as well as its comparison to some state of the art benchmark controller. Owing to the relaxed barrier function being convex and twice continuously differentiable, we can formulate an efficient convex optimization problem to solve the aforementioned optimization problems, see Section 4.2. We provide a suitable *numerical* optimization technique to solve the convex optimization problem. Furthermore, we test the different algorithms with respect to different maneuvers and performance criteria in the area of an automated driving vehicle. Finally, we apply the proposed algorithm for trajectory tracking to a real-world vehicle and show experimental results.

In Chapter 5, we give a short conclusion and summary as well as an outline of future research areas.

1.3 Background

Set point stabilization, trajectory tracking, and path following: *Set-point stabilization,* e.g., the stabilization of a pendulum in its inverted position, deals with the design of a feedback controller such that the error between the desired set point and the closed-loop trajectory asymptotically converges to zero, stays small for all times or at least stays within certain bounds around the desired set point. The design of such a feedback controller is well-known in control theory and was already mechanically performed in the 1780s with the invention of the flyball governor by James Watt, see J. Maxwell (1868). However, there is a large problem class subsumed under the term *trajectory tracking,* in which we want to track a time-varying reference in the state or output space such as the temperature profile of a chemical reactor or the guidance of a tactic missle towards a moving target. This problem class is by definition not covered by the set point stabilization problem. While in a set point stabilization problem, we want to track a desired time-invariant set point, the objective in trajectory tracking is to track a time-varying reference. Trajectory tracking is already more challenging than set-point stabilization, see G. Hoffmann and S. Waslander (2008) and J. Yang and J. Kim (1999), and suitable for automated driving, due to the fact that we like to follow a time-dependent reference. However, if we think of an autonomous vehicle that follows a geometric path with a desired velocity profile, the main objective is to converge to the path geometrically. At second interest, we want to follow the path with the desired velocity. Thus, the typical objective in automated driving differs from the standard trajectory tracking problem. This leads us to the problem class of *path following,* which is even more useful for automated driving, although more difficult. Compared to trajectory tracking, the timing of the reference trajectory is not given a priori. Thus, not only the control feedback but also the time evolution along the geometric path has to be chosen online by the feedback controller, which can be seen as an additional degree of freedom for the controller. The advantage of the path following controller lies in the possibility to adapt the time evolution of the reference trajectory online, which directly corresponds to the adaptation of the desired velocity profile. Since the 1990s, the problem class of path following received wide attention in the area of robotics, see C. Canudas and R. Roskam (1991); P. Chiacchio (1990). Since then, many works on the path following problem class were published, see A. Aguiar and J. Hespanha (2007); T. Faulwasser (2012).

MPC: There are many methods of solving set-point stabilization, trajectory tracking, and path following problems such as linear, nonlinear, geometric, nongeometric, model-based, and model-free approaches, to mention only a few. However, in almost any practical application, the dynamical system has to satisfy certain constraints. MPC has turned out to be an effective method of solving constrained control tasks. It refers to a model-based control method in which, at each sampling instant, an open-loop optimal control problem is solved to obtain a suitable control input. The solution to the open-loop optimal control problem and the overall closed-loop structure of MPC are shown in Figure 1.2 below.

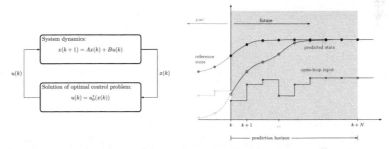

Figure 1.2. Left: Overall closed-loop structure of an MPC scheme. The first element of the solution $u_0^*(x(k))$ of the open-loop optimal control problem is applied to a linear time-invariant discrete-time system with system state $x(k)$ and control input $u(k)$ at sampling instant $k \in \mathbb{N}$. Right: The solution of the open-loop optimal control problem consists of a predictive open-loop input sequence $\{u(k), \ldots, u(k+N-1)\}$ which drives the corresponding predicted state sequence towards the reference state.

In particular, we consider linear time-invariant discrete-time systems

$$x(k+1) = Ax(k) + Bu(k), \quad x(0) = x_0, \tag{1.1}$$

where $x(k) \in \mathbb{R}^n$, $u(k) \in \mathbb{R}^m$ denote the system state and the control input at sampling instant $k \in \mathbb{N}$. We assume (A, B) to be stabilizable. Furthermore, we impose that the state and input have to satisfy the following polytopic state and input constraints

$$x(k) \in \mathcal{X} = \{x \in \mathbb{R}^n : C_x x \leq d_x\} \qquad u(k) \in \mathcal{U} = \{u \in \mathbb{R}^m : C_u u \leq d_u\}, \tag{1.2}$$

for all $k \in \mathbb{N}$ where $C_x \in \mathbb{R}^{q_x \times n}$, $d_x \in \mathbb{R}^{q_x}_{++}$, $C_u \in \mathbb{R}^{q_u \times m}$, $d_u \in \mathbb{R}^{q_u}_{++}$, and $q_x, q_u \in \mathbb{N}$ denote the number of polytopic state and input constraints. In the following, we will explain the basic ideas of a stabilizing MPC scheme that guarantees constraint satisfaction at the example of a set-point stabilization problem. Basically, the control goal in set-point stabilization is to stabilize a desired set point, which can, without the loss of generality, be assumed to be the origin $(x_s, u_s) = (0,0)$. Next, we will explain the concept of terminal set-based MPC for linear time-invariant discrete-time systems. The corresponding open-loop optimal control problem is defined as

$$
\begin{aligned}
J_N^*(x(k)) = \sum_{i=0}^{N-1} &\ell(x_i, u_i) + F(x_N) \\
\text{s.t. } x_{i+1} &= Ax_i + Bu_i, \quad x_0 = x(k) \\
x_i &\in \mathcal{X}, \quad i = 0, \ldots, N, \\
u_i &\in \mathcal{U}, \quad i = 0, \ldots, N-1, \\
x_N &\in \mathcal{X}_f,
\end{aligned}
\tag{1.3}
$$

where $x(k)$ denotes the current measured system state at sampling instant $k \in \mathbb{N}$, $N \in \mathbb{N}_+$ refers to the prediction horizon, and $\mathcal{X}_f \subseteq \mathbb{R}^n$ denotes the terminal set. The stage cost $\ell : \mathbb{R}^n \times \mathbb{R}^m \to \mathbb{R}_{++}$ and the terminal cost $F : \mathbb{R}^n \to \mathbb{R}_+$ are defined as

$$
\ell(x_i, u_i) = ||x_i||_Q^2 + ||u_i||_R^2, \qquad F(x_N) = ||x_N||_P^2,
\tag{1.4}
$$

where $Q \in \mathbb{S}_{++}^n$, $R \in \mathbb{S}_{++}^m$ and $P \in \mathbb{S}_+^n$. The resulting feedback of the corresponding MPC scheme is obtained from the solution of the open-loop optimal control problem parametrized by the current measured system state at sampling instant $k \in \mathbb{N}$. In particular, the solution of the open-loop optimal control problem is an optimal input sequence denoted as $u^*(x(k)) = \{u_0^*(x(k))\ u_1^*(x(k)) \ldots u_{N-1}^*(x(k))\}$. As usual, only the first element of the input sequence $u^*(x(k))$ is applied to the system. Thus, the closed-loop arises to

$$
x(k+1) = Ax(k) + Bu_0^*(x(k)).
\tag{1.5}
$$

While the state and input cost matrices Q and R are typically chosen by the user, the terminal state weight matrix P and the terminal set $\mathcal{X}_f \subseteq \mathbb{R}^n$ have to be chosen in a specific way such that the origin is rendered asymptotically stable and the optimization problem remains feasible for all sampling instants $k \in \mathbb{N}$, see J. Rawlings, D. Mayne and M. Diehl (2019). In particular, we choose the terminal state weight matrix P to be the solution of the discrete-time algebraic

Riccati equation

$$P = (A + BK)^\top P(A + BK) + K^\top RK + Q$$
$$K = -(R + B^\top PB)^{-1} B^\top PA. \tag{1.6}$$

Assuming that the system (A, B) is stabilizable and $Q \in \mathbb{S}^n_{++}$ and $R \in \mathbb{S}^m_{++}$, there exists a solution (P, K) to (1.6) with $P \in \mathbb{S}^n_{++}$ and with the control gain K being stabilizing, see J. Rawlings, D. Mayne and M. Diehl (2019). Hence, we are able to choose the terminal set as $\mathcal{X}_f = \{x \in \mathbb{R}^n : x(i) = (A + BK)^i x \subset \mathcal{X}, u(i) = Kx(i) \in \mathcal{U} \ \forall i \geq 0\}$, see E. Gilbert and T. Tan (1991). Assuming that the MPC scheme associated with the open-loop optimal control is initially feasible, we are always able to construct a feasible solution by the terminal set $\mathcal{X}_f$ because of K being a stabilizing control gain. Furthermore, the MPC scheme associated with the open-loop optimal control problem (1.3) asymptotically stabilizes the origin of the closed-loop system (1.5). This can be shown by using the cost function $J^*_N(x(k))$ as a Lyapunov function, which will be important for the remainder of the thesis. However, conventional MPC schemes suffer from high computational effort and infeasibility of the open-loop optimal control problem in the presence of model uncertainties or disturbances. One recent and interesting approach to overcome these disadvantages is the so-called rbMPC, see C. Feller (2017); C. Feller and C. Ebenbauer (2016, 2017, 2020).

Relaxed barrier function MPC: As already pointed out, standard MPC schemes generally suffer from two main drawbacks. On the one hand, the mismatch between the prediction model of the open-loop optimal control problem and the actual system may lead to infeasibility of the open-loop optimal control problem. On the other hand, MPC schemes have to be implemented on some hardware where computation power is restricted to a hard limit. Hence, the application of suboptimal inputs (a solution which is not fully iterated) is beneficial for the MPC scheme to guarantee hard realtime requirements. To overcome the drawbacks of conventional MPC schemes, we introduce the notion of *rbMPC*. A highly detailed description of the theoretical framework can be found in C. Feller (2017). However, we want to revise the most important facts. We start by introducing the notion of barrier functions according to C. Feller (2017); C. Feller and C. Ebenbauer (2016, 2017, 2020). Consider the most simple inequality constrained optimization problem with

$$\min_{z \in \mathbb{R}} \ f(z)$$
$$\text{s.t.} \ -z \leq 0, \tag{1.7}$$

where $f : \mathbb{R} \to \mathbb{R}$ is a scalar function. It is assumed that $f \in \mathcal{C}^2$ and f being strictly convex where $\mathcal{C}^2$ belongs to the set of twice continuously differentiable functions. The inequality constrained optimization problem (1.7) can be transformed into the following unconstrained optimization problem

$$\min_{z \in \mathbb{R}} f(z) + I(z) \tag{1.8}$$

with the ideal barrier function, also known as indicator function, $I : \mathbb{R} \to \mathbb{R}_+$ defined as

$$I(z) = \begin{cases} 0, & z \geq 0 \\ \infty, & z < 0. \end{cases} \tag{1.9}$$

Hence, barrier functions handle inequality constraints by including them in the cost function of the optimization problem. Obviously, the ideal barrier function I is non-continuous at $z = 0$ and thus it is numerically not tractable. This motivates us to approximate the ideal barrier function by means of the logarithmic barrier function $B : \mathbb{R}_{++} \to \mathbb{R}$ with

$$B(z) = -\ln(z). \tag{1.10}$$

Obviously, $B(z) \in \mathcal{C}^2$ for $z \in (0, \infty]$, $B(z)$ is convex and $B(z) \to \infty$ as $z \to 0$. The unconstrained optimization problem (1.8) can be rewritten into

$$\min_{z \in \mathbb{R}} f(z) + \epsilon B(z), \qquad \epsilon \in \mathbb{R}_{++}. \tag{1.11}$$

A graphical illustration of the logarithmic barrier function and the indicator function is given in Figure 1.3. From Figure 1.3, it is obvious that the smaller the weighting parameter ϵ gets, the better the weighted logarithmic barrier function $\epsilon B(z)$ approximates the indicator function $I(z)$. Applying the Karush-Kuhn-Tucker (KKT) conditions to (1.11), we have that

$$\nabla f(z^*(\epsilon)) + \frac{\epsilon}{z^*(\epsilon)} = 0, \tag{1.12}$$

where $z^*(\epsilon) \in \mathbb{R}$ is the optimal solution to the unconstrained barrier function based optimization problem (1.11). If we introduce $\lambda^*(\epsilon) = \frac{\epsilon}{-z^*(\epsilon)}$, then the KKT conditions (1.12), see C. Feller (2017), transform into

$$\begin{aligned} \nabla f(z^*(\epsilon)) + \lambda^*(\epsilon) &= 0, \\ -\lambda^*(\epsilon)z^*(\epsilon) &= \epsilon, \\ \lambda^*(\epsilon) \geq 0, \quad -z^*(\epsilon) &\leq 0. \end{aligned} \tag{1.13}$$

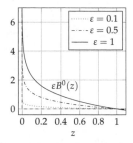

Figure 1.3. The figure shows the logarithmic barrier function $B^0(z)$ for different values of $\epsilon = 0.1$ (⋯⋯), $\epsilon = 0.5$ (-⋅-⋅), $\epsilon = 1$ (———) and the indicator function $I(z)$ (- - -). Note that for $z > 1$, the logarithmic barrier function becomes negative.

By applying Lagrangian duality concepts, see S. Boyd and L. Vandenberghe (2004), we get an upper bound on the suboptimality of the approximated problem (1.11), which is

$$f(z^*(\epsilon)) - p^* \leq m\epsilon, \tag{1.14}$$

where $p^* \in \mathbb{R}$ refers to the optimal solution of the constrained optimization problem (1.7) and $m \in \mathbb{N}_+$ denotes the number of inequality constraints (in our case, $m = 1$), see S. Boyd and L. Vandenberghe (2004). In conventional MPC stability analysis, the optimal value function $J^*(z)$ is used as a Lyapunov function where the minimum is attained at the origin. This requires that the value function is positive definite. For the unconstrained optimization problem (1.11), neither the mimimum is attained at the origin nor is the logarithmic barrier non-negative, see Figure 1.4 and 1.3. To overcome these issues, we introduce the notion of *recentered barrier* functions, see A. Wills and W. Heath (2002).

Definition 1. Let $B^0 : \mathcal{D} \to \mathbb{R}$ be a barrier function on an open and non-empty convex set $\mathcal{D} \subseteq \mathbb{R}$, which contains the origin in its interior. Then, the function $B_G^0 : \mathcal{D} \to \mathbb{R}_+$, defined as

$$B_G^0(z) - B^0(z) - B^0(0) - \left[\nabla B^0(0)\right]^\top z \tag{1.15}$$

is the gradient recentered barrier function for the set $\mathcal{D}$ around the origin, see C. Feller (2017)[Def. 2.5, p.33].

A graphical illustration of the recentered barrier function is given in Figure 1.4.

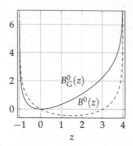

Figure 1.4. The figure shows the recentered logarithmic barrier function $B_G^0(z)$ (——) and the non-recentered logarithmic barrier function $B^0(z)$ (- - -) for $z \in \mathcal{Z} = \{z \in \mathbb{R} : -1 \leq z \leq 4\}$.

Obviously, the use of barrier functions transforms an inequality constraint optimization problem like (1.7) into a numerically tractable unconstrained optimization problem like (1.11), which is *me* suboptimal with respect to the original problem. However, if we think of the model mismatch between the actual system behavior and the prediction model used in the open-loop optimal control problem, we know that it will cause a certain prediction error. Hence, it might happen at the limit of the imposed constraints that there exist values $z < 0$. Owing to the definition of the logarithmic barrier function (1.10), it leads to an infeasible optimization problem. This motivates the usage of *relaxed barrier functions*, first proposed in C. Feller and C. Ebenbauer (2016). The relaxed barrier functions are defined as

$$\hat{B}^0(z) = \begin{cases} B^0(z), & z > \delta \\ \beta(z; \delta), & z \leq \delta, \end{cases} \tag{1.16}$$

where $\beta(\cdot; \delta) : \mathbb{R} \to \mathbb{R}$ is the relaxation function and $\delta \in \mathbb{R}_{++}$ refers to the relaxation parameter. The relaxation function $\beta(\cdot; \cdot)$ is chosen such that the logarithmic barrier function is smoothly extended for $z \leq \delta$. As proposed in C. Feller (2017)[p. 45], we make use of the following quadratic barrier function

$$\beta(z; \delta) = \frac{1}{2}\left[\left(\frac{z - 2\delta}{\delta}\right)^2 - 1\right] - \ln(\delta). \tag{1.17}$$

A graphical illustration of the relaxed barrier function is given in Figure 1.5. Hence, $\hat{B}^0(z) \in \mathcal{C}^2$ for $z \in [-\infty, \infty]$, $\hat{B}^0(z)$ is convex and $\hat{B}^0(z) \to \infty$ as $z \to -\infty$. Compared to the definition of barrier functions (1.3), the relaxed barrier functions are globally defined because of its relaxation. The corresponding relaxed

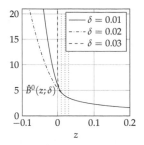

Figure 1.5. The figure shows the relaxed barrier function $\hat{B}^0(z)$ for different values of $\delta = 0.01$ (——), $\delta = 0.02$ (–·–·) and $\delta = 0.03$ (– – –) for $z \in \mathcal{Z} = \{z \in \mathbb{R} : 0 \leq z\}$.

gradient recentered barrier function around the origin is defined as

$$\hat{B}_G^0(z) = \begin{cases} B_G^0(z), & z > \delta \\ \beta(z; \delta), & z \leq \delta. \end{cases} \tag{1.18}$$

In the following, we shortly recap the advantages of relaxed barrier functions. Firstly, the usage of the relaxed barrier function transfers the inequality constraints into the cost. Thus, we obtain, depending on the choice of the decision variables, an equality constrained or even an unconstrained convex optimization problem that can be solved efficiently. Secondly, the relaxed barrier functions are globally defined. Hence, the open-loop optimal control problem will be always feasible.

In this chapter, we have introduced the reader to the problem setup of automated driving. From a control and system theoretical point of view, the problem setup of automated driving belongs both to the *trajectory tracking* problem and the more suitable but even more complicated *path following* problem. Given that almost every system is restricted to physical constraints, we pointed out that MPC is an appealing method of solving the tracking problem and simultaneously satisfying the imposed constraints. However, we have concluded that standard MPC schemes suffer from certain disadvantages that motivated us to use a special form of MPC, the so-called rbMPC in which the inequality constraints are transformed into the cost by using relaxed barrier functions. The following chapters will deal with the application of rbMPC algorithms to the trajectory tracking and path following problem for linear time-invariant discrete-time systems.

2

Tracking and path following in model predictive control

This chapter consists of two parts. The first part considers the trajectory tracking problem and the second part deals with the path following problem. Solving both problems enable us to follow a time-varying reference in the state space. In the remainder, we aim to solve the trajectory tracking and path following problem with an MPC formulation based on relaxed barrier functions.

In trajectory tracking or reference tracking, the control goal is to track a time-varying state (or output) reference trajectory such that the tracking error remains small or asymptotically converges to zero. The tracking error is defined as the deviation between the actual and the desired trajectory for the actual sampling instant. Standard trajectory tracking applications are, e.g., automated driving or positional tracking of an industrial servo machine tool drive. Trajectory tracking problems solved by means of MPC have been addressed in several publications, see for example A. Aguiar and J. Hespanha (2007); A. Aguiar, J. Hespanha and P. Kokotovic (2005); D. Limon et al. (2008); D. Simon, J. Löfberg and T. Glad (2014); U. Maeder and M. Morari (2010). In path following, the goal is to follow a parametrized geometric path, see A. Aguiar and J. Hespanha (2007); A. Aguiar, J. Hespanha and P. Kokotovic (2005); T. Fossen et al. (2003), or parametrized trajectories in the state (or output) space, see T. Faulwasser (2012). Similar to trajectory tracking, the goal is that the path following error remains small or asymptotically converges to zero. Compared to trajectory tracking, the timing of the reference trajectory is not given a priori. Thus, not only the control feedback but also the time evolution along the parametrized curve has to be controlled online by the feedback controller. Thus, the controller has the ability to adapt the time evolution of the reference path which corresponds to the adaption of the desired velocity profile. This can be seen as an additional degree of freedom for the controller to overcome certain performance limitations, see

A. Aguiar, J. Hespanha and P. Kokotovic (2005).

The main part of this chapter deals with the theoretical investigation of stability and constraint satisfaction properties of a trajectory tracking and path following rbMPC scheme for linear time-invariant discrete-time systems. We show that a purely terminal cost-based rbMPC tracking and path following scheme achieves exponential stability of the reference trajectory and reference path for the closed-loop system, respectively. Moreover, we derive conditions that guarantee exact constraint satisfaction for a certain set of initial conditions. Finally, we show simulation results that confirm our theoretical investigations for both problem classes. The results presented in this chapter are based on F. Pfitz, X. Hu and C. Ebenbauer (2021) and F. Pfitz, C. Ebenbauer and M. Braun (2019).

2.1 Problem setup

Consider a discrete-time linear time-invariant system

$$x(k+1) = Ax(k) + Bu(k), \qquad x(0) = x_0, \tag{2.1}$$

where $x(k) \in \mathbb{R}^n$ and $u(k) \in \mathbb{R}^m$ refer to the system state and control input at sampling instant $k \in \mathbb{N}$, respectively. The following polytopic state and input constraints

$$\begin{aligned} x(k) &\in \mathcal{X} = \{x \in \mathbb{R}^n : C_x x \leq d_x\}, \\ u(k) &\in \mathcal{U} = \{u \in \mathbb{R}^m : C_u u \leq d_u\} \end{aligned} \tag{2.2}$$

are imposed where $C_x \in \mathbb{R}^{q_x \times n}$, $C_u \in \mathbb{R}^{q_u \times m}$, $d_x \in \mathbb{R}^{q_x}_{++}$, and $d_u \in \mathbb{R}^{q_u}_{++}$. Here, $q_x, q_u \in \mathbb{N}$ represent the number of state and input constraints.

Assumption 2. The system (A, B) is stabilizable.

As already pointed out in the introduction of this chapter, trajectory tracking and path following differ in the definition of the reference. In trajectory tracking the reference corresponds to a, in general, time-varying trajectory in the state or output space. Whereas in path following, the reference is characterized by a parametrized path in the state or output space without knowing the timing a priori. In the following, we consider each problem separately. We start with investigating the trajectory tracking problem, and continue with the analysis of the path following problem. In all cases, we will restrict ourselves to references in the state and input space.

2.1.1 Trajectory tracking for linear discrete-time systems

In this section, we address the trajectory tracking problem based on rbMPC. First, we define the reference trajectory which we like to follow. Second, we state the specifications of the trajectory tracking controller. Finally, we formulate the open-loop optimal control problem based on relaxed barrier functions such that the control goal is fulfilled.

Assumption 3. We assume that the reference trajectory $r : \mathbb{N} \to \mathbb{R}^n \times \mathbb{R}^m$ is given by

$$r(k) = (x_{\mathrm{ref}}(k), u_{\mathrm{ref}}(k)), \tag{2.3}$$

a priori, where $x_{\mathrm{ref}}(k) \in \mathbb{R}^n$ and $u_{\mathrm{ref}}(k) \in \mathbb{R}^m$ denote the reference system state and reference control input at sampling instant k, respectively. We assume that the reference trajectory satisfies the following properties: **a)** it strictly lies in the polytopic state and input constraints (2.2), i.e., $x_{\mathrm{ref}}(k) \in \mathcal{X}_{\mathrm{ref}} = \{x_{\mathrm{ref}} \in \mathbb{R}^n : C_x x_{\mathrm{ref}} \leq \bar{d}_x, \bar{d}_x = d_x - \epsilon_x\}, \epsilon_x \in \mathbb{R}^{q_x}_{++}$ and $u_{\mathrm{ref}}(k) \in \mathcal{U}_{\mathrm{ref}} = \{u_{\mathrm{ref}} : C_u u_{\mathrm{ref}} \leq \bar{d}_u, \bar{d}_u = d_u - \epsilon_u\}, \epsilon_u \in \mathbb{R}^{q_u}_{++}$ for all $k \in \mathbb{N}$, **b)** it is a solution for the system dynamics (2.1), i.e.,

$$x_{\mathrm{ref}}(k+1) = A x_{\mathrm{ref}}(k) + B u_{\mathrm{ref}}(k), \qquad x_{\mathrm{ref}}(0) = x_{\mathrm{ref},0}. \tag{2.4}$$

The goal is to design an rbMPC trajectory tracking scheme, that achieves

1. **Asymptotic stability** of the closed-loop system (2.18) for the origin which results in trajectory convergence. In particular, we like to achieve that the system state error converges asymptotically to zero

$$\lim_{k \to \infty} ||x(k) - x_{\mathrm{ref}}(k)|| = 0. \tag{2.5}$$

2. **Constraint satisfaction**, in which the polytopic state and input constraints (2.2) have to be satisfied for all $k \in \mathbb{N}$.

In the following, we formulate the MPC problem for trajectory tracking based on relaxed barrier functions. In order to use the relaxed barrier functions from Definition 2 which are recentered around the origin, we rewrite the constraints (2.2) in error coordinates. More precisely, we have

$$\begin{aligned} x(k) - x_{\mathrm{ref}}(k) \in \mathcal{E}_x(k) &= \{(e_x, k) \in \mathbb{R}^n \times \mathbb{N} : C_x e_x \leq d_x(k)\}, \\ u(k) - u_{\mathrm{ref}}(k) \in \mathcal{E}_u(k) &= \{(e_u, k) \in \mathbb{R}^m \times \mathbb{N} : C_u e_u \leq d_u(k)\}, \end{aligned} \tag{2.6}$$

where $d_x(k) = d_x - C_x x_{ref}(k) \in \mathbb{R}_{++}^{q_x}$ and $d_u(k) = d_u - C_u u_{ref}(k) \in \mathbb{R}_{++}^{q_u}$ for any $k \in \mathbb{N}$, due to Assumption 3 a.). In a general MPC scheme, an open-loop optimal control problem is solved at every sampling instance $k \in \mathbb{N}$. For the given trajectory tracking problem, the open-loop optimal control problem can be formulated as

$$J_N^*(x(k), k) = \min_{\boldsymbol{u}} \sum_{j=0}^{N-1} \ell(x_j, u_j, k+j) + F(x_N, k+N)$$

$$\text{s.t. } x_{j+1} = A x_j + B u_j, \quad x_0 = x(k), \tag{2.7}$$

$$x_j - x_{ref}(k+j) \in \mathcal{E}_x(k+j), \quad j = 0, \dots, N-1,$$

$$u_j - u_{ref}(k+j) \in \mathcal{E}_u(k+j), \quad j = 0, \dots, N-1,$$

where $x(k) \in \mathbb{R}^n$ refers to the current measured system state and $\{x_{ref}(k+j)\}_{j=0,\dots,N}$, $\{u_{ref}(k+j))\}_{j=0,\dots,N-1}$ at sampling instant $k \in \mathbb{N}$ denote the respective state and input reference along the prediction horizon $N \in \mathbb{N}_+$. Furthermore, $\boldsymbol{u} = \{u_0, \dots, u_{N-1}\}$ specifies the decision variables. The optimal open-loop input sequence over the prediction horizon is denoted as $u^*(x(k), k) = \{u_0^*(x(k), k), \dots, u_{N-1}^*(x(k), k)\}$. Note that the subindices x_j, u_j refer to the open-loop predictions of the open-loop optimal control problem and $x(k)$, $u(k)$ denote the actual state and input trajectory. The stage cost $\ell : \mathbb{R}^n \times \mathbb{R}^m \times \mathbb{N} \to \mathbb{R}_+$ is defined as

$$\ell(x_j, u_j, k+j) = ||x_j - x_{ref}(k+j)||_Q^2 + ||u_j - u_{ref}(k+j)||_R^2, \tag{2.8}$$

with $Q \in \mathbb{S}_{++}^n$ and $R \in \mathbb{S}_+^m$ being suitably chosen tuning matrices. The terminal cost $F : \mathbb{R}^n \times \mathbb{N} \to \mathbb{R}_+$ is denoted as

$$F(x_N, k+N) = ||x_N - x_{ref}(k+N)||_P^2, \tag{2.9}$$

where $P \in \mathbb{S}_{++}$ is designed such that stability of the proposed MPC scheme is guaranteed, see Theorem 1. As usual, the resulting feedback of the proposed MPC scheme is obtained from the solution of the open-loop optimal control problem based on the current system state $x(k)$ and the state and input reference trajectories $\{x_{ref}(k+j)\}_{j=0,\dots,N}$, $\{u_{ref}(k+j))\}_{j=0,\dots,N-1}$ along the prediction horizon. In particular, only the first element of the optimal input sequence $u^*(x(k), k) = \{u_0^*(x(k), k), \dots, u_{N-1}^*(x(k), k)\}$ is applied to the system dynamics. The resulting closed-loop system arises to

$$x(k+1) = A x(k) + B u_0^*(x(k), k), \qquad x(0) = x_0. \tag{2.10}$$

Next, the open-loop optimal control problem (2.7) is reformulated using relaxed barrier functions. As already pointed out in Chapter 1, the basic idea is to eliminate the constraints and to introduce relaxed barrier functions as penalty terms into the cost functional, see C. Feller and C. Ebenbauer (2017). Because of the use of relaxed barrier functions, recursive feasibility and the so-called anytime property can be guaranteed by design under rather mild assumptions, see C. Feller (2017) for the set-point stabilization problem. It is worthwhile to point out that, despite the use of relaxed barrier functions, exact constraint satisfaction can be guaranteed, see Subsection 2.2.1 Theorem 2. For the generalization of the rbMPC to the problem setup of trajectory tracking, we introduce the following definitions.

Definition 1. Let $B(z) = -\ln(z)$ be the logarithmic barrier function corresponding to the set $\mathbb{R}_+$. Moreover, consider a scalar $\delta \in \mathbb{R}_{++}$, called the relaxation parameter, and let $\beta(z;\delta) : (-\infty, \delta] \to \mathbb{R}$ be a strictly monotone and continuously differentiable function that satisfies $\beta(\delta;\delta) = -\ln(\delta)$ as well as $\lim_{z \to -\infty} \beta(z;\delta) = \infty$. Then, we call $\hat{B} : \mathbb{R}_+ \to \mathbb{R}$ defined as

$$\hat{B}(z) = \begin{cases} B(z) & z > \delta \\ \beta(z;\delta) & z \le \delta \end{cases} \tag{2.11}$$

the *relaxed logarithmic barrier function* and refer to the function $\beta(z;\delta)$ as the *relaxing function*. In particular, we will make use of the following polynomial relaxing function

$$\beta_1(z;\delta) = \frac{l-1}{l}\left[\left(\frac{z-l\delta}{(l-1)\delta}\right)^l - 1\right] - \ln(\delta), \tag{2.12}$$

where $l > 1$ is an even integer. In the remainder, we will choose $l = 2$ as suggested by C. Feller (2017) and J. Hauser and A. Saccon (2006). ●

Definition 2. Let $\mathcal{D}$ be an open and non-empty convex set that contains the origin and let $\hat{B} : \mathcal{D} \to \mathbb{R}_+$ be a relaxed barrier function according to Definition 1. Then the function $\hat{B}_G : \mathcal{D} \to \mathbb{R}_+$ with

$$\hat{B}_G(z) = \hat{B}(z) - \hat{B}(0) - [\nabla\hat{B}(0)]^\top z \tag{2.13}$$

is called the *relaxed gradient recentered* logarithmic barrier function of the set $\mathcal{D}$ around the origin. ●

Definition 3. Let $\mathcal{P}(k) = \{z \in \mathbb{R}^r : Cz \leq d(k)\}$ with $C \in \mathbb{R}^{q \times r}$ and $d : \mathbb{N} \to \mathbb{R}_{++}^q$ be a time-varying polytopic set that contains the origin for all $k \in \mathbb{N}$. Moreover, let $0 < \delta \leq \min(d(k))$ for all $k \in \mathbb{N}$. Then,

$$\hat{B}_G(z,k) = \sum_{i=1}^{q} \hat{B}(-C^i z + d^i(k)) + \ln(d^i(k)) - \frac{C^i z}{d^i(k)} \tag{2.14}$$

defines the *time-varying gradient recentered relaxed logarithmic barrier function* for the polytopic set $\mathcal{P}(k)$ around the origin. •

It is important to mention that relaxing and recentering the logarithmic barrier function is mandatory to achieve global feasibility and positive definiteness. A graphical illustration of the time-invariant case of Definition 3 is given in Section 1.3, see Figure 1.5. With the help of Definition 3, the open-loop optimal control problem (2.7) of the rbMPC tracking scheme can be formulated as

$$\hat{J}_N^*(x(k),k) = \min_u \sum_{j=0}^{N-1} \ell(x_j, u_j, k+j) + F(x_N, k+N) \tag{2.15}$$

$$\text{s.t. } x_{j+1} = Ax_j + Bu_j, \quad x_0 = x(k),$$

where the stage cost $\ell : \mathbb{R}^n \times \mathbb{R}^m \times \mathbb{N} \to \mathbb{R}_{++}$ is defined as

$$\ell(x_j, u_j, k+j) := ||x_j - x_{\text{ref}}(k+j)||_Q^2 + ||u_j - u_{\text{ref}}(k+j)||_R^2 + \epsilon \hat{B}_G(x_j, u_j, k+j) \tag{2.16}$$

with $\epsilon \in \mathbb{R}_{++}$, where $\hat{B}_G(x_j, u_j, k+j) = \hat{B}_{G,x}(x_j - x_{\text{ref}}(k+j), k+j) + \hat{B}_{G,u}(u_j - u_{\text{ref}}(k+j), k+j)$ for all $k, j \in \mathbb{N}$ contains the gradient recentered relaxed logarithmic barrier function, see Definition 3, for the time-varying polytopic error sets (2.6). Similar to eq. (2.10) the closed loop arises to

$$x(k+1) = Ax(k) + B\hat{u}_0^*(x(k),k), \quad x(0) = x_0. \tag{2.17}$$

where $\hat{u}_0^*(x(k),k)$ corresponds to the first element of the input sequence of the open-loop optimal control problem (2.15). The corresponding closed-loop error system results in

$$e_x(k+1) = Ae_x(k) + Be_u(k), \quad e_{x,0} = e_x(0) \tag{2.18}$$

with the state and input error $e_x(k) = x(k) - x_{\text{ref}}(k)$ and $e_u(k) = \hat{u}_0^*(x(k),k) - u_{\text{ref}}(k)$ at sampling instant $k \in \mathbb{N}$, respectively.

2.1.2 Path following for linear discrete-time systems

In the following, we examine the path following problem for an MPC scheme based on relaxed barrier functions. Similar to the previous section, we first give a definition of the reference path, we then state the control goals for the path following controller and finally we provide a formulation of an MPC controller based on relaxed barrier function which fulfills the defined control goals. As already pointed out at the beginning of Chapter 2, the path following problem is a more natural but also more complex formulation for automated driving. It provides an additional degree of freedom to the controller design which has certain advantages compared to trajectory tracking, see A. Aguiar, J. Hespanha and P. Kokotovic (2005). In the following, we start with several assumptions.

Assumption 4. We assume that the reference path $P : \mathbb{R} \to \mathbb{R}^n \times \mathbb{R}^m$ that has to be followed is given by

$$P(\theta) = (x_{\text{ref}}(\theta), u_{\text{ref}}(\theta)) \tag{2.19}$$

where $x_{\text{ref}}(\theta) \in \mathbb{R}^n$ and $u_{\text{ref}}(\theta) \in \mathbb{R}^m$ denote the reference system state and reference control input parametrized with the path parameter $\theta \in \mathbb{R}$.

Assumption 5. The path (2.19) has the following properties: **a)** it strictly lies within the polytopic state and input constraint set, meaning that $x_{\text{ref}}(\theta) \in \mathcal{X}_{\text{ref}}$ and $u_{\text{ref}}(\theta) \in \mathcal{U}_{\text{ref}}$ for all $\theta \in \mathbb{R}$ i.e., $x_{\text{ref}}(\theta) \in \mathcal{X}_{\text{ref}} = \{x_{\text{ref}} \in \mathbb{R}^n : C_x x_{\text{ref}} \leq \tilde{d}_x, \tilde{d}_x = d_x - \epsilon_x\}, \epsilon_x \in \mathbb{R}_{++}^{q_x}$ and $u_{\text{ref}}(\theta) \in \mathcal{U}_{\text{ref}} = \{u_{\text{ref}} : C_u u_{\text{ref}} \leq \tilde{d}_u, \tilde{d}_u = d_u - \epsilon_u\}, \epsilon_u \in \mathbb{R}_{++}^{q_u}$ for all $\theta \in \mathbb{R}$, **b)** it is consistent with the system dynamics (2.1) for $\theta = k$, i.e., $x_{\text{ref}}(k+1) = Ax_{\text{ref}}(k) + Bu_{\text{ref}}(k), x_{\text{ref}}(0) = x_{\text{ref},0}$ with $x_{\text{ref}}(k) \in \mathcal{X}_{\text{ref}}$ and $u_{\text{ref}}(k) \in \mathcal{U}_{\text{ref}}$ for any $k \in \mathbb{N}$.

Remark 6. In general, the path variable θ is associated with a time dependent update law $\theta = \theta(k)$ for all $k \in \mathbb{N}$ which is not specified a priori. Determining $\theta(k)$ generates an additional degree of freedom, see T. Faulwasser and R. Findeisen (2011), which can be seen as a reparametrization of time, see I. Batkovic et al. (2020). In this thesis, we consider parametrized curves in the state and input space, see eq. (2.19).

Having defined the reference path, we like to design an rbMPC path following scheme, that achieves

1. **Asymptotic stability** of the closed-loop system for the reference path which results in path convergence. In particular, we like to achieve that the system state error converges asymptotically to zero

$$\lim_{k \to \infty} ||x(k) - x_{\text{ref}}(\theta(k))|| = 0, \tag{2.20}$$

2. **Strict forward motion**, in which the system (2.1) moves forward along the pre-described path (2.19), meaning that $\theta(k+1) > \theta(k)$ with $\theta(0) = \theta_0$, where the initial state θ_0 of the update law (2.23) is obtained by solving the following minimum distance problem

$$\theta_0 = \arg\min_{\theta \in \mathbb{R}} ||x_0 - x_{\text{ref}}(\theta)||^2, \tag{2.21}$$

see J. Hauser and R. Hindmann (1995); T. Faulwasser (2012),

3. **Constraint satisfaction**, in which the polytopic state and input constraints (2.53) and (2.52) have to be satisfied for all $k \in \mathbb{N}$.

Remark 7. Similar to Subsection 2.1.1, we transform the constraints (2.2) into time-varying polytopic error state and error input constraints. More precisely, we have

$$\begin{aligned}
x(k) - x_{\text{ref}}(\theta) &\in \mathcal{E}_{\text{x}}(\theta) = \{(e_{\text{x}}, \theta) \in \mathbb{R}^n \times \mathbb{R} : C_x e_x \leq d_{\text{x}}(\theta)\} \\
u(k) - u_{\text{ref}}(\theta) &\in \mathcal{E}_{\text{u}}(\theta) = \{(e_{\text{u}}, \theta) \in \mathbb{R}^m \times \mathbb{R} : C_u e_u \leq d_{\text{u}}(\theta)\}
\end{aligned} \tag{2.22}$$

for any $\theta \in \mathbb{R}$ with $d_{\text{x}}(\theta) \in \mathbb{R}^{q_x}_{++}$ and $d_u(\theta) \in \mathbb{R}^{q_u}_{++}$, due to Assumption 5. A natural choice to ensure strict forward motion is

$$\theta(k+1) = \theta(k) + 1 + v(k), \qquad \theta(0) = \theta_0, \tag{2.23}$$

where $v(k) \in \mathcal{V} = \{v \in \mathbb{R} : v \geq -1 - \epsilon_{\text{v}}, \epsilon_{\text{v}} > 0\}$ can be seen as an additional virtual control input at sampling instant $k \in \mathbb{N}$. The set $\mathcal{V}$ can be reformulated into a polytopic constraint set such that

$$v(k) \in \mathcal{V} = \{v \in \mathbb{R} : C_v v \leq d_{\text{v}} - \epsilon_{\text{v}}\}, \ \epsilon_{\text{v}} \in \mathbb{R}_{++} \tag{2.24}$$

with suitably chosen $C_v \in \mathbb{R}$ and $d_{\text{v}} - \epsilon_{\text{v}} \in \mathbb{R}_{++}$. Note that, if $v(k) \equiv 0$, the update law (2.23) simplifies to $\theta(k+1) = \theta(k) + 1$ for any $k \in \mathbb{N}$, and the problem task turns into a trajectory tracking problem, see F. Pfitz, C. Ebenbauer and M. Braun (2019).

First, we state the open-loop optimal control without the usage of relaxed barrier functions which we call standard MPC scheme. Second, we transform the standard MPC scheme into an rbMPC scheme. For the given path following problem, the open-loop optimal control problem, associated with a standard

MPC scheme, can be formulated as

$$J_N^*(x(k), \theta(k)) = \min_{u,v} \sum_{j=0}^{N-1} \ell(x_j, u_j, \theta_j) + F(x_N, \theta_N)$$

$$\begin{aligned}
\text{s.t. } & x_{j+1} = Ax_j + Bu_j, \quad x_0 = x(k), \\
& \theta_{j+1} = \theta_j + 1 + v_j, \quad \theta_0 = \theta(k), \\
& x_j - x_{\text{ref}}(\theta_j) \in \mathcal{E}_x(\theta_j), \ j = 0, \dots, N-1, \\
& u_j - u_{\text{ref}}(\theta_j) \in \mathcal{E}_u(\theta_j), \ j = 0, \dots, N-1, \\
& v_j \in \mathcal{V}, \ j = 0, \dots, N-1,
\end{aligned} \tag{2.25}$$

where $x(k)$ denotes the current system state and $\theta(k)$ refers to the path parameter at sampling instant $k \in \mathbb{N}$, respectively. Similar to Subsection 2.1.1, we have that $u = \{u_0, \dots, u_{N-1}\}$ and $v = \{v_0, \dots, v_{N-1}\}$ denote the control input and virtual control input decision variables of the open-loop optimal control problem. Note that the subindices x_j, u_j, θ_j, v_j are used for the predictions of the system state, control input, path parameter, and virtual control input, while $x(k), u(k), \theta(k)$, and $v(k)$ refer to the actual realizations. The stage cost $\ell : \mathbb{R}^n \times \mathbb{R}^m \times \mathbb{R} \to \mathbb{R}_{++}$ is defined as

$$\ell(x_j, u_j, \theta_j) = ||x_j - x_{\text{ref}}(\theta_j)||_Q^2 + ||u_j - u_{\text{ref}}(\theta_j)||_R^2, \tag{2.26}$$

where $Q \in \mathbb{S}_{++}^n$ and $R \in \mathbb{S}_{++}^m$ are user-defined tuning matrices. The terminal cost is $F(x_N, \theta_N) = ||x_N - x_{\text{ref}}(\theta_N)||_P^2$ with $P \in \mathbb{S}_{++}^n$, which will be designed in Section 2.2.2 such that stability of the reference path (2.19) for the closed-loop system is guaranteed. The solution of the open-loop optimal control problem defined in (2.25) is denoted as $u^*(x(k), \theta(k)) = \{u_0^*(x(k), \theta(k)) \ u_1^*(x(k), \theta(k)) \ \dots \ u_{N-1}^*(x(k), \theta(k))\}$ and $v^*(x(k), \theta(k)) = \{v_0^*(x(k), \theta(k)) \ v_1^*(x(k), \theta(k)) \ \dots \ v_{N-1}^*(x(k), \theta(k))\}$. Only the first element of the input sequence $u^*(x(k), \theta(k))$ and the virtual input sequence $v^*(x(k), \theta(k))$ are applied to system (2.1). Thus, the closed-loop system is given by

$$x(k+1) = Ax(k) + Bu_0^*(x(k), \theta(k)), \quad x(0) = x_0, \tag{2.27a}$$
$$\theta(k+1) = \theta(k) + 1 + v_0^*(x(k), \theta(k)), \quad \theta(0) = \theta_0. \tag{2.27b}$$

In the following, we formulate the open-loop optimal control problem (2.25) by using relaxed barrier functions. As already pointed out in Chapter 1, again the basic idea is to eliminate the constraints by writing the constraints in terms of relaxed barrier functions as penalty terms into the cost. In the following, we make use of Definitions 1 and 2 to introduce the reader to gradient recentered barrier functions for the path following problem.

Definition 4. Let $P(\theta) = \{z \in \mathbb{R}^r : Cz \leq d(\theta)\}$ with $C \in \mathbb{R}^{q \times r}$ and $d : \mathbb{R} \to \mathbb{R}^q_{++}$ be a polytopic set that contains the origin for all $\theta \in \mathbb{R}$. Furthermore, let $P = \{z \in \mathbb{R}^r : Cz \leq d\}$ with $d \in \mathbb{R}^q_{++}$ be also a polytopic set that contains the origin. Moreover, let $0 \leq \delta \leq \min(d(\theta))$ for all $\theta \in \mathbb{R}$ with $\delta \in \mathbb{R}_{++}$. Then, the function

$$\hat{B}_G(z, \theta) = \sum_{i=1}^{q} \hat{B}(-C^i z + d^i(\theta)) + \ln(d^i(\theta)) - \frac{C^i z}{d^i(\theta)} \tag{2.28}$$

defines the *gradient recentered logarithmic barrier function* for the *polytopic set* $P(\theta)$ and the function

$$\hat{B}_G(z) = \sum_{i=1}^{q} \hat{B}(-C^i z + d^i) + \ln(d^i) - \frac{C^i z}{d^i} \tag{2.29}$$

defines the *gradient recentered relaxed logarithmic barrier function* for the *polytopic set* P. •

Note that, due to Assumption 5 a.) there always exists a $\delta \in \mathbb{R}_{++}$. Using Definition 4, the open-loop optimal control problem (2.25) of the rbMPC path following scheme can be formulated as

$$\hat{J}_N^*(x(k), \theta(k)) = \min_{\mathbf{u}, \mathbf{v}} \sum_{j=0}^{N-1} \ell(x_j, u_j, v_j, \theta_j) + F(x_N, \theta_N)$$

$$\text{s.t. } x_{j+1} = Ax_j + Bu_j, \quad x_0 = x(k), \tag{2.30}$$

$$\theta_{j+1} = \theta_j + 1 + v_j, \quad \theta_0 = \theta(k).$$

The stage cost $\ell : \mathbb{R}^n \times \mathbb{R}^m \times \mathbb{R} \times \mathbb{R} \to \mathbb{R}_+$ is defined as

$$\ell(x_j, u_j, v_j, \theta_j) = ||x_j - x_{\text{ref}}(\theta_j)||^2_Q + ||u_j - u_{\text{ref}}(\theta_j)||^2_R$$

$$+ \varepsilon \hat{B}_{G,xv}(x_j, u_j, \theta_j, v_j), \tag{2.31}$$

with $\hat{B}_{G,xv}(x_j, u_j, \theta_j, v_j) = \hat{B}_{G,x}(x_j - x_{\text{ref}}(\theta_j), \theta_j) + \hat{B}_{G,u}(u_j - u_{\text{ref}}(\theta_j), \theta_j) + \hat{B}_{G,v}(v_j)$ and weighting parameter $\varepsilon \in \mathbb{R}_{++}$ with the gradient recentered relaxed barrier functions $\hat{B}_{G,x} : \mathbb{R}^n \times \mathbb{R} \to \mathbb{R}_+$, $\hat{B}_{G,u} : \mathbb{R}^m \times \mathbb{R} \to \mathbb{R}_+$, and $B_{G,v} : \mathbb{R} \to \mathbb{R}_+$. Similar to eq. (2.27a) - (2.27b) the closed loop arises to

$$x(k+1) = Ax(k) + B\hat{u}_0^*(x(k), \theta(k)), \quad x(0) = x_0, \tag{2.32}$$

$$\theta(k+1) = \theta(k) + 1 + \hat{v}_0^*(x(k), \theta(k)), \quad \theta(0) = \theta_0, \tag{2.33}$$

where $\hat{u}_0^*(x(k), \theta(k))$ and $\hat{v}_0^*(x(k), \theta(k))$ correspond to the first element of the input sequence and the virtual input sequence of the open-loop optimal control problem (2.30).

The following section deals with investigations of stability of the closed-loop systems (2.17) and (2.32) - (2.33) for both, the proposed trajectory tracking controller and path following controller based on relaxed barrier functions. As usual, the choice of the terminal cost $F(x_N, \theta_N)$ plays a decisive role in stability properties. Different realizations of a standard stabilizing relaxed barrier function approach are given in C. Feller (2017), we will restrict ourselves to a terminal-set free approach purely based on a quadratic terminal cost.

2.2 Closed-loop stability and constraint satisfaction

2.2.1 Trajectory tracking: Closed-loop stability and exact constraint satisfaction

For the stability proof, it will be important that the gradient recentered relaxed barrier function can be quadratically upper bounded time invariantly for all $k \in \mathbb{N}$. Thus, we start with several Lemmatas about the properties of relaxed barrier functions.

Lemma 1. Let $\hat{B}_G : \mathbb{R}^r \times \mathbb{N} \to \mathbb{R}_+$ be a time-varying gradient recentered logarithmic barrier function with respect to the polytopic set $\mathcal{P}(k) = \{z \in \mathbb{R}^r : Cz \leq d(k), \}$ with $d(k) \in \mathbb{R}_{++}$ for all $k \in \mathbb{N}$ which contains the origin. Let the underlying relaxing function be quadratic as in eq. (2.12) and $0 < \delta \leq d(k)$ for all $k \in \mathbb{N}$ be given. Then, it holds that

$$\hat{B}_G(z, k) \leq z^\top M_r z, \tag{2.34}$$

where $M_r \in \mathbb{S}_{++}^r$ is defined as $M_r = \frac{1}{2\delta^2} C^\top C$. $\quad\bullet$

For a proof, we refer the reader to Appendix B.1.

Lemma 2. Let $\hat{B}_K : \mathbb{R}^n \times \mathbb{N} \to \mathbb{R}_+$ be the time-varying relaxed gradient recentered barrier function with respect to the state and input error polytopes (2.6). Moreover, let the relaxing function be quadratic and $\delta \in \mathbb{R}_{++}$ be given. Then, for any $K \in \mathbb{R}^{n \times m}$, the gradient recentered and relaxed barrier function $\hat{B}_K(x - x_{\text{ref}}(k), k) = \hat{B}_{G,x}(x - x_{\text{ref}}(k), k) + \hat{B}_{G,u}(K(x - x_{\text{ref}}(k)), k)$ satisfies

$$\hat{B}_K(x - x_{\text{ref}}(k), k) \leq (x - x_{\text{ref}}(k))^\top M(x - x_{\text{ref}}(k)), \tag{2.35}$$

for all $(x, k) \in \mathbb{R}^n \times \mathbb{N}$ with $M = M_x + K^\top M_u K$, where $M_x = \frac{1}{2\delta^2} C_x^\top C_x$ and $M_u = \frac{1}{2\delta^2} C_u^\top C_u$. $\quad\bullet$

For a proof, we refer the reader to Appendix B.2. Note that (2.35) is a time-invariant upper bound of the respective time-varying relaxed barrier function. In the following, we investigate the stability properties of the closed-loop system (2.17) for the reference trajectory defined in Assumption 3. We like to mention that we use Lemma 10 given in Appendix A to prove that the reference is globally exponential stable for the closed-loop system (2.17).

Theorem 1. Let Assumptions 2 – 3 be satisfied. Then, the closed-loop system (2.18) is globally uniformly exponentially stable for the origin. •

The proof is given in Appendix B.3 and basically uses a modified discrete-time algebraic Riccati equation as well as the upper bound from Lemma 2. Next, we investigate strict constraint satisfaction of the rbMPC scheme for the trajectory tracking problem. Motivated by C. Feller (2017), we start with several definitions.

Definition 5. We define the scalar $\bar{\beta}(\delta) \in \mathbb{R}_{++}$ as

$$\bar{\beta}(\delta) := \min\{\bar{\beta}_x(\delta), \bar{\beta}_u(\delta)\}, \tag{2.36}$$

with

$$\bar{\beta}_x(\delta) = \min_{e_x,k}\{\hat{B}_{G,x}(e_x,k)|C_x e_x \leq d_x(k), C_x^i e_x = d_x^i(k), i = 1, \ldots, q_x, k \in \mathbb{N}\},$$

$$\bar{\beta}_u(\delta) = \min_{e_u,k}\{\hat{B}_{G,u}(e_u,k)|C_u e_u \leq d_u(k), C_u^i e_u = d_u^i(k), i = 1, \ldots, q_u, k \in \mathbb{N}\},$$

$$\tag{2.37}$$

where $\bar{\beta}(\delta)$ is a lower bound on the minimal values of the state and input barrier functions that can be achieved on the respective borders of the constraint sets $\mathcal{E}_x(k)$ and $\mathcal{E}_u(k)$ for all $k \in \mathbb{N}$, see (2.6). •

In order to ensure that the minimum in Definition 5 exists, we make the following additional assumption on the reference trajectory.

Assumption 8. We assume that the reference trajectory, see Assumption 3, ends in a steady-state of (2.4), i.e., $x_{\text{ref}}(k+1) = x_{\text{ref}}(k)$ for all $k \geq \bar{k}$ with $\bar{k} \in \mathbb{N}$ or moves along a periodic reference $x_{\text{ref}}(k) = x_{\text{ref}}(k+T)$ with period $T \in \mathbb{R}$.

Lemma 3. Let $\bar{\beta}_x(\delta)$ and $\bar{\beta}_u(\delta)$ be defined according to Definition 5. Then, the associated barrier function sublevel sets satisfy

$$\mathcal{S}_x(\delta) := \{(e_x,k) \in \mathbb{R}^n \times \mathbb{N}|\hat{B}_{G,x}(e_x,k) \leq \bar{\beta}_x(\delta)\} \subseteq \mathcal{E}_x(k), \tag{2.38a}$$

$$\mathcal{S}_u(\delta) := \{(e_u,k) \in \mathbb{R}^m \times \mathbb{N}|\hat{B}_{G,u}(e_u,k) \leq \bar{\beta}_u(\delta)\} \subseteq \mathcal{E}_u(k), \tag{2.38b}$$

for all $k \in \mathbb{N}$ which means that the sublevel sets (2.38a) - (2.38b) will always be contained within the time-varying error constraint sets $\mathcal{E}_x(k)$, $\mathcal{E}_u(k)$. •

For a proof, we follow along the ideas of (C. Feller, 2017, C.2) and (C. Feller, 2017, Lemma 3.1). We will outline the proof for the state sublevel set $\mathcal{S}_x(\delta)$. Note that the same arguments hold for the input sublevel set. We will prove Lemma 3 by contradiction. Assume there exists a $\bar{e}_x \in \mathcal{S}_x(\delta)$ with $\bar{e}_x \notin \mathcal{E}_x(k)$ and a $e_x \in \mathcal{S}_x(\delta)$ with $e_x \in \mathcal{E}_x(k)$ for some $k \in \mathbb{N}$. Due to $e_x, \bar{e}_x \in \mathcal{S}_x(\delta)$, we can conclude that $\hat{B}_{G,x}(e_x, k) \le \bar{\beta}_x(\delta)$ and $\hat{B}_{G,x}(\bar{e}_x, k) \le \bar{\beta}_x(\delta)$ for all $k \in \mathbb{N}$. Due to $\hat{B}_G(\bar{e}_x, k)$ and $\hat{B}_G(e_x, k)$ being bounded for all $k \in \mathbb{N}$ and being strictly monotonic increasing, there exists for every $k \in \mathbb{N}$ a $\lambda \in [0, 1)$ such that $e_x = \lambda \bar{e}_x$ for

$$\hat{B}_{G,x}(e_x, k) = \hat{B}_{G,x}(\lambda \bar{e}_x, k). \tag{2.39}$$

Due to convexity of $\hat{B}_{G,x}(\cdot, k)$ in the first argument, we have that

$$\hat{B}_{G,x}(\lambda \bar{e}_x + (1 - \lambda)\tilde{e}_x, k) \le \lambda \hat{B}_{G,x}(\bar{e}_x, k) + (1 - \lambda)\hat{B}_{G,x}(\tilde{e}_x, k) \le \lambda \bar{\beta}(\delta) \tag{2.40}$$

for $\lambda \in [0, 1)$ and any $k \in \mathbb{N}$. If we choose $\tilde{e}_x$ such that $\hat{B}_{G,x}(\tilde{e}_x, k) = 0$ which is due to the definition of the gradient recentered barrier functions $\tilde{e}_x = 0$, we result in

$$\hat{B}_{G,x}(e_x, k) = \hat{B}_{G,x}(\lambda \bar{e}_x, k) \le \lambda \bar{\beta}(\delta) \tag{2.41}$$

for some $\lambda \in [0, 1)$ and for some $k \in \mathbb{N}$. Due to the definition of $\bar{\beta}(\delta)$ being the minimum, this cannot hold true and thus $\bar{e}_x \in \mathcal{S}_x(\delta)$ with $\bar{e}_x \notin \mathcal{E}_x(k)$ cannot exist and the opposite must hold. $\qquad \square$

Definition 6. Let the optimal value function of the *unconstrained infinite-horizon problem* be defined as

$$J_\infty^*(x(k), k) = \min_{\boldsymbol{u}} \ \sum_{j=0}^{\infty} \ell_\infty(x_j, u_j, k + j) \tag{2.42}$$
$$\text{s.t.} \ x_{j+1} = Ax_j + Bu_j, \quad x_0 = x(k),$$

where $x(k) \in \mathbb{R}^n$ refers to the current measured system state at sampling instant $k \in \mathbb{N}$ and decision variables $\boldsymbol{u} = \{u_0, \dots, u_{N-1}\}$. The stage cost $\ell_\infty : \mathbb{R}^n \times \mathbb{R}^m \times \mathbb{N} \to \mathbb{R}_{++}$ is defined as

$$\ell_\infty(x_j, u_j, k + j) = ||x_j - x_{\text{ref}}(k + j)||_Q^2 + ||u_j - u_{\text{ref}}(k + j)||_R^2. \tag{2.43}$$

with $Q \in \mathbb{S}_{++}^n$ and $R \in \mathbb{S}_{++}^m$. $\qquad \bullet$

Theorem 2. Let Assumptions 2, 3 and 8 be satisfied and let $\hat{\mathcal{E}}_N(\delta) = \{x \in \mathbb{R}^n | \hat{J}_N^*(x, 0) - J_\infty^*(x, 0) \le \bar{\beta}(\delta)\}$. For any $x(0) \in \hat{\mathcal{E}}_N(\delta)$ the state and input constraints are strictly satisfied for the closed-loop system (2.17) for any $k \in \mathbb{N}$. $\qquad \bullet$

The proof is given in Appendix B.4. The basic idea is to use the sublevel sets defined in (2.38a) and (2.38b) to show that the system state and control input $x(k)$ and $u(k)$ will satisfy initially the polytopic input and state error sets (2.6). We then use the stability results to show that the system state and control input trajectory will stay within the sublevel sets at all times and thus the point-wise state and input constraints are satisfied for all $k \in \mathbb{N}$. Note that an analytic solution of $J_\infty^*(x, 0) = ||x - x_{\text{ref}}(0)||_P^2$ for all $x \in \mathbb{R}^n$ exists which originates from the discrete-time Bellman equation for the infinite horizon problem.

2.2.2 Path following: Closed-loop stability and exact constraint satisfaction

In this subsection, we like to examine stability and exact constraint satisfaction properties of an rbMPC path following controller. In the following, we exploit the fact that the gradient recentered relaxed barrier function can be quadratically and invariantly upper bounded for all $\theta \in \mathbb{R}$. We start our investigations with several Lemmatas about the properties of relaxed barrier functions tailored for the path following problem.

Lemma 4. Let $\hat{B}_G : \mathbb{R}^r \times \mathbb{R} \to \mathbb{R}_+$ be a time-varying gradient recentered logarithmic barrier function with respect to the polytopic set $P(\theta) = \{z \in \mathbb{R}^r : Cz \leq d(\theta)\}$ with $C \in \mathbb{R}^{q \times r}$, $d(\theta) \in \mathbb{R}_{++}^q$ which contains the origin. Let the underlying relaxing function be quadratic, see eq. (2.12) for $i = 2$, chosen with a suitable relaxation parameter $\delta \in \mathbb{R}_{++}$ with $0 \leq \delta \leq \min(d(\theta))$ for any $\theta \in \mathbb{R}$. Then, it holds that

$$\hat{B}_G(z, \theta) \leq z^\top M_r z, \tag{2.44}$$

where $M_r = \frac{1}{2\delta^2} C^\top C \in \mathbb{S}_{++}^r$. ●

The proof is given in Appendix B.5. The Lemma basically exploits the properties of the relaxed barriers, see Definition 4, to show that there exists a uniform upper bound of the relaxed barrier function for all $\theta \in \mathbb{R}$.

Lemma 5. Let $\hat{B}_{G,x} : \mathbb{R}^n \times \mathbb{R} \to \mathbb{R}_+$ and $\hat{B}_{G,u} : \mathbb{R}^m \times \mathbb{R} \to \mathbb{R}_+$ be gradient recentered logarithmic barrier functions according to Definition 4. Then, for any control gain $K \in \mathbb{R}^{m \times n}$, the associated barrier function $\hat{B}_K(x - x_{\text{ref}}(\theta), \theta) = \hat{B}_{G,x}(x - x_{\text{ref}}(\theta), \theta) + \hat{B}_{G,u}(K(x - x_{\text{ref}}(\theta)), \theta)$ satisfies

$$\hat{B}_K(x - x_{\text{ref}}(\theta), \theta) \leq (x - x_{\text{ref}}(\theta))^\top M (x - x_{\text{ref}}(\theta)) \tag{2.45}$$

for all $(x, \theta) \in \mathbb{R}^n \times \mathbb{R}$ with $M = M_x + K^\top M_u K$ and $M_x \in \mathbb{S}_+^n$, $M_u \in \mathbb{S}_+^m$ with $M_x = \frac{1}{2\delta} C_x^\top C_x$ and $M_u = \frac{1}{2\delta} C_u^\top C_u$. ●

The proof is given in the Appendix B.6. Motivated by C. Feller (2017), we choose the terminal cost matrix $P \in \mathbb{S}_{++}^n$ to be the positive definite solution of a modified discrete-time Riccati equation $P = A_K^{\top} P A_K + K^{\top}(R + \varepsilon M_u)K + Q + \varepsilon M_x$. Note that a stabilizing control gain $K = -(R + B^{\top} PB + \varepsilon M_u)^{-1} B^{\top} PA$ always exists, given that Q and R are positive definite and the system (A, B) is assumed to be stabilizable, see J. Rawlings, D. Mayne and M. Diehl (2019). We summarize the finding in the following Assumption.

Assumption 9. There exists $Q \in \mathbb{S}_{++}^n$, $R \in \mathbb{S}_{++}^m$, $\varepsilon \in \mathbb{R}_{++}$ and a $K \in \mathbb{R}^{m \times n}$ such that $A_K = A + BK$ is Schur stable.

Next, we start our stability analysis by using a partial stability result based on semidefinite Lyapunov functions, see Lemma 11 in Appendix A.

Theorem 3. Let Assumptions 2 and $4 - 9$ hold. Then, the closed-loop system (2.32) - (2.33) is globally exponentially stable uniformly in $\theta \in \mathbb{R}$ for the reference path $P(\theta)$. •

For a rigorous proof, we refer the reader to Appendix B.7. Next, we derive a special set of initial conditions, in which constraint satisfaction of the rbMPC scheme path following problem can be guaranteed. We start with the following definitions, similar to C. Feller (2017).

Definition 7. We define the scalar $\bar{\beta}(\delta) \in \mathbb{R}_{++}$ as

$$\bar{\beta}(\delta) := \min\{\bar{\beta}_x(\delta), \bar{\beta}_u(\delta), \bar{\beta}_v(\delta)\}, \tag{2.46}$$

where

$$\bar{\beta}_x(\delta) = \min_{e_x, \theta}\{\hat{B}_{G,x}(e_x, \theta) | C_x e_x \leq d_x(\theta), C_x^i e_x = d_x^i(\theta), i = 1, \ldots, q_x, \theta \in \mathbb{R}\},$$
$$\bar{\beta}_u(\delta) = \min_{e_u, \theta}\{\hat{B}_{G,u}(e_u, \theta) | C_u e_u \leq d_u(\theta), C_u^i e_u = d_u^i(\theta), i = 1, \ldots, q_u, \theta \in \mathbb{R}\},$$
$$\bar{\beta}_v(\delta) = \min_v\{\hat{B}_{G,v}(v) | C_v v \leq d_v, C_v = d_v\}.$$

$$\tag{2.47}$$

•

In order to ensure that the minimum in Definition 7 exists, we make the following additional assumption on the reference path.

Assumption 10. We assume that the reference path, see Assumption 4, ends in a steady-state of (2.1), meaning that $x_{\text{ref}}(\theta) = x_{\text{ref}}(\tilde{\theta})$ for all $\theta \geq \tilde{\theta}$ with $\tilde{\theta} \in \mathbb{R}$ or moves along some periodic reference, i.e. $x_{\text{ref}}(\theta) = x_{\text{ref}}(\theta + T)$ with period $T \in \mathbb{R}$.

Note that $\bar{\beta}(\delta)$ is a lower bound on the minimal values of the barrier functions $\hat{B}_{G,x}(\cdot,\theta)$, $\hat{B}_{G,u}(\cdot,\theta)$ and $\hat{B}_{G,v}(\cdot)$ that can be obtained on the respective borders of the sets $\mathcal{E}_x(\theta)$, $\mathcal{E}_u(\theta)$, and $\mathcal{V}$ for any $\theta \in \mathbb{R}$. We continue with the definition of the barrier function sublevel sets which is decisive for ensuring strict constraint satisfaction under a special set of initial conditions.

Lemma 6. Let $\bar{\beta}_x(\delta)$, $\bar{\beta}_u(\delta)$ and $\bar{\beta}_v(\delta)$ be defined according to Definition 7. Then, the associated barrier function sublevel sets

$$\mathcal{S}_{\hat{B}_x}(\delta) = \{(e_x,\theta) \in \mathbb{R}^n \times \mathbb{R}| \hat{B}_{G,x}(e_x,\theta) \leq \bar{\beta}_x(\delta)\}, \tag{2.48a}$$

$$\mathcal{S}_{\hat{B}_u}(\delta) = \{(e_u,\theta) \in \mathbb{R}^m \times \mathbb{R}| \hat{B}_{G,u}(e_u,\theta) \leq \bar{\beta}_u(\delta)\}, \tag{2.48b}$$

$$\mathcal{S}_{\hat{B}_v}(\delta) = \{v \in \mathbb{R}| \hat{B}_{G,v}(v) \leq \bar{\beta}_v(\delta)\}, \tag{2.48c}$$

that satisfy $\mathcal{S}_{\hat{B}_x}(\delta) \subseteq \mathcal{E}_x(\theta)$, $\mathcal{S}_{\hat{B}_u}(\delta) \subseteq \mathcal{E}_u(\theta)$ and $\mathcal{S}_{\hat{B}_v}(\delta) \subseteq \mathcal{V}$ for any $\theta \in \mathbb{R}$. •

Note that similar arguments as in case of the trajectory tracking, see Lemma 6, hold.

Definition 8. Let the optimal value function of the *unconstrained infinite-horizon problem* be defined as

$$J_\infty^*(x(k),\theta(k)) = \min_{u,v} \sum_{j=0}^{\infty} \ell_\infty(x_j,u_j,\theta_j)$$
$$\text{s.t. } x_{j+1} = Ax_j + Bu_j, \quad x_0 = x(k), \tag{2.49}$$
$$\theta_{j+1} = \theta_j + 1 + v_j, \quad \theta_0 = \theta(k),$$

where $x(k) \in \mathbb{R}^n$ refers to the system state and $\theta(k) \in \mathbb{R}$ denotes the path parameter at sampling instant $k \in \mathbb{N}$, respectively. The decision variables are defined as $u = \{u_i\}_{i=0}^{\infty}$ and $v = \{v_i\}_{i=0}^{\infty}$. Moreover, the stage cost $\ell_\infty : \mathbb{R}^n \times \mathbb{R}^m \times \mathbb{R} \to \mathbb{R}_{++}$ is given by

$$\ell_\infty(x_j,u_j,\theta_j) = ||x_j - x_{\text{ref}}(\theta_j)||_Q^2 + ||u_j - u_{\text{ref}}(\theta_j)||_R^2. \tag{2.50}$$

•

Theorem 4. Let Assumptions 2, 4 – 10 hold and let $\hat{\mathcal{E}}_N(\delta,\theta(0)) = \{x \in \mathbb{R}^n|$ $\hat{J}_N^*(x,\theta(0)) - J_\infty^*(x,\theta(0)) \leq \varepsilon\bar{\beta}(\delta)\}$ for some $\theta(0) \in \mathbb{R}$. Then, for any $x(0) \in \hat{\mathcal{E}}_N(\delta,\theta(0))$, the state, input, and virtual input constraints are strictly satisfied for the closed-loop system (2.32) – (2.33) for any $k \in \mathbb{N}$. •

The proof of Theorem 4 is given in Appendix B.8. Similar to the trajectory tracking problem, the basic idea is to use the sublevel sets in Definition 6 to show that the system state $x(k)$, control input $u(k)$ and virtual control input $v(k)$ will satisfy initially the polytopic input, virtual input and system state error sets. We then use the stability results from Theorem 3 to prove that the system state, control input and virtual control input trajectory will stay within the sublevel sets for all times and thus the point-wise state, input and virtual constraints are satisfied for all $k \in \mathbb{N}$.

2.3 Numerical example

In this section, we like to verify our stability results and exact constraint satisfaction guarantees for the rbMPC trajectory tracking and rbMPC path following algorithm. We start our investigations with a discrete-time double integrator for the rbMPC trajectory tracking algorithm. The system dynamics of the discrete-time double integrator are given by

$$x(k+1) = \begin{bmatrix} 1 & h \\ 0 & 1 \end{bmatrix} x(k) + \begin{bmatrix} \frac{h^2}{2} \\ h \end{bmatrix} u(k), \quad x(0) = x_0, \tag{2.51}$$

where $h = 0.1$ refers to the step size. Note that (2.51) is obtained by exact discretization of its continuous-time dynamics. Furthermore, we impose the following state and input constraints

$$\mathcal{U} = \{u \in \mathbb{R} : -14 \leq u \leq 10\}, \tag{2.52}$$

$$\mathcal{X} = \{x \in \mathbb{R}^2 : -1 \leq x_1 \leq 1, -2.5 \leq x_2 \leq 2.5\}. \tag{2.53}$$

The relaxation parameter and the weighting parameter of the relaxed barrier functions are chosen as $\epsilon = 1e^{-1}$ and $\delta = 1e^{-5}$. The reference trajectory satisfies the discrete-time system dynamics of (2.51) and strictly stays within the state and input constraint set $\mathcal{X}$ and $\mathcal{U}$. Hence, Assumptions 3 a.) and b.) are fulfilled. The prediction horizon of the open-loop optimal control problem is selected to be $N = 10$ and the weighting matrices are given by $Q = \begin{bmatrix} 10 & 0 \\ 0 & 10 \end{bmatrix}$ and $R = 1$.

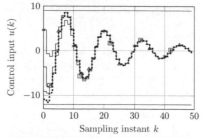

a) Control input $u(k)$ of the closed-loop system for different initial conditions $x(0) = x_{0,i}$, $i = 1, 2, 3$ depicted as (—∗—), (—□—) and (—▲—). The reference control input is displayed as (······).

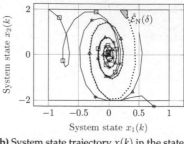

b) System state trajectory $x(k)$ in the state space of the closed-loop system for different initial conditions $x(0) = x_{0,i}$, $i = 1, 2, 3$ depicted as (—∗—), (—□—), (—▲—), the reference trajectory is shown as (······). Moreover, the set of exact constraint satisfaction $\hat{\mathcal{E}}_N(\delta)$ is displayed.

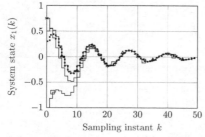

c) System state trajectory $x_1(k)$ of the closed-loop system for different initial conditions $x(0) = x_{0,i}$, $i = 1, 2, 3$ depicted as (—∗—), (—□—), (—▲—). The reference trajectory is shown as (······).

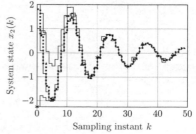

d) System state trajectory $x_2(k)$ of the closed-loop system for different initial conditions $x(0) = x_{0,i}$, $i = 1, 2, 3$ depicted as (—∗—), (—□—), (—▲—). The reference trajectory is shown as (······).

Figures 2.1a - 2.1e show the actual realization of the control input $u(k)$, the system state $x(k)$ and the optimal value function $J_N^*(x(k), k)$ of the closed-loop system (2.17) with respect to different initial conditions $x(0) = x_{0,i}$, $i = 1, 2, 3$ with $x_{0,1} = \begin{bmatrix} 0.75 & -2.25 \end{bmatrix}^\top$, $x_{0,2} = \begin{bmatrix} -1 & 2 \end{bmatrix}^\top$, and $x_{0,3} = \begin{bmatrix} 0.5 & 1 \end{bmatrix}^\top$. From Figures 2.1a - 2.1e, we can conclude that the system state $x(k)$ of the closed-loop system (2.17) is converging towards the reference trajectory for all initial condi-

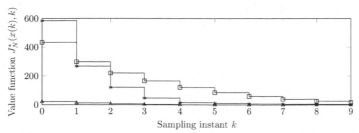

e) Optimal value function $J_N^*(x(k), k)$ of the closed-loop system for different initial conditions $x(0) = x_{0,i}$, $i = 1, 2, 3$ depicted as (—*—), (—□—) and (—▲—).

Figure 2.1. Simulation results of the discrete-time double integrator (2.51) for the proposed trajectory tracking rbMPC controller.

tions $x_{0,i=1...3}$. Due to Theorem 2, we can infer that if $x(0) \in \hat{E}_N(\delta)$, then exact state and input satisfaction is guaranteed. However, we see that if $x(0)$ does not initially lie within the set $\hat{\mathcal{E}}_N(\delta)$, see $x(0) = x_{0,1/2/3}$, then exact state and input constraint satisfaction is still possible. This fact originates from the conservative bound used within the calculation of the set $\hat{\mathcal{E}}_N(\delta)$. Moreover, we can conclude that if the initial condition lies outside of the system state constraint set $x(0) = x_{0,3} \notin \mathcal{X}$, the rbMPC trajectory tracking algorithm is still able to calculate a stabilizing control input. This characteristic originates from the usage of a relaxed barrier function and its global definition. Finally, we like to mention that once the system state and control input entered the constraint sets $\mathcal{X}$ and $\mathcal{U}$, the system state trajectory $x(k)$ and control input trajectory $u(k)$ remain within their respective bounds.

Remark 11. We calculated the set $\hat{\mathcal{E}}_N(\delta)$ by discretizing the state constraint set $\mathcal{X}$ with step size 0.01. According to Theorem 2, we checked if the current state belongs to the set $\hat{\mathcal{E}}_N(\delta)$.

We continue our stability and exact constraint satisfaction verification for the rbMPC path following algorithm. We examine the example of an actuated undamped oscillator. The system dynamics of the discrete-time actuated undamped oscillator are denoted as

$$x(k+1) = \begin{bmatrix} 0.995 & 0.0998 \\ -0.998 & 0.995 \end{bmatrix} x(k) + \begin{bmatrix} 0.005 \\ 0.0998 \end{bmatrix} u(k), \quad x(0) = x_0, \quad (2.54)$$

where (2.54) refers to the exact discretization of the continuous-time dynamics $\ddot{x} + x = u$ of an undamped oscillator with step size $h = 0.1$. We impose the following state, input and virtual input constraints

$$\mathcal{U} = \{u \in \mathbb{R} : -1 \leq u \leq 1\}, \ \mathcal{V} = \{v \in \mathbb{R} : -0.9 \leq v\}, \tag{2.55}$$

$$\mathcal{X} = \{x \in \mathbb{R}^2 : -1.25 \leq x_1 \leq 1.25, -1.25 \leq x_2 \leq 1.25\}. \tag{2.56}$$

The following Figures 2.2a - 2.2f show the actual realization of the control input $u(k)$, the system state $x(k)$ and the optimal value function $J_N^*(x(k), \theta(k))$ of the closed-loop system (2.17) with respect to different initial conditions $x(0) = x_{0,i}, i = 1, 2, 3$ with $x_{0,1} = \begin{bmatrix} 0 & -1.5 \end{bmatrix}^\top, x_{0,2} = \begin{bmatrix} 1 & -1 \end{bmatrix}^\top$, and $x_{0,3} = \begin{bmatrix} -0.5 & 0.5 \end{bmatrix}^\top$. The relaxation parameter and the weighting parameter of the relaxed barrier functions are chosen as $\epsilon = 1e^{-3}$ and $\delta = 1e^{-4}$. Furthermore, we choose the reference path to be $P(\theta) = (x_1(\theta) = \cos(\theta h), x_2(\theta) = -\sin(\theta h), u(\theta) = 0)$ for all $\theta \in \mathbb{R}$. Hence, the reference path satisfies the discrete-time system dynamics of (2.54) for $\theta = k$ and strictly stays within the state and input constraint set $\mathcal{X}$ and $\mathcal{U}$. Thus, Assumptions 5 a.) and b.) are fulfilled. The prediction horizon of the open-loop optimal control problem is selected to be $N = 10$ and the weighting matrices are given by $Q = \begin{bmatrix} 10 & 0 \\ 0 & 10 \end{bmatrix}$ and $R = 1$.

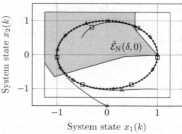

a) Control input $u(k)$ of the closed-loop system for different initial conditions $x(0) = x_{0,i}, i = 1, 2, 3$ depicted as (─✱─), (─▭─) and (─▲─). The reference control input is displayed as (┄┄┄).

b) System state trajectory $x(k)$ in the state space of the closed-loop system for different initial conditions $x(0) = x_{0,i}, i = 1, 2, 3$ depicted as (─✱─), (─▭─), (─▲─), the reference trajectory is shown as (┄┄┄). Moreover, the set of exact constraint satisfaction $\hat{\mathcal{E}}_N(\delta, 0)$ is displayed.

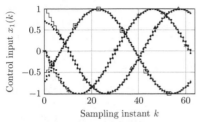
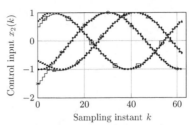

c) System state trajectory $x_1(k)$ of the closed-loop system for different initial conditions $x(0) = x_{0,i}, i = 1,2,3$ depicted as (—✳—), (—▫—), (—▲—) , the reference trajectory is shown as (·······).

d) System state trajectory $x_2(k)$ of the closed-loop system for different initial conditions $x(0) = x_{0,i}, i = 1,2,3$ depicted as (—✳—), (—▫—), (—▲—) , the reference trajectory is shown as (·······).

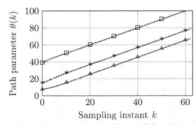
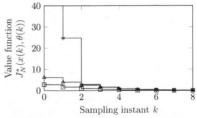

e) Path parameter update $\theta(k)$ of the closed-loop system for different initial conditions $x(0) = x_{0,i}, i = 1,2,3$ depicted as (—✳—), (—▫—), (—▲—) .

f) Optimal value function $J_N^*(x(k),,\theta(k))$ of the closed-loop system for different initial conditions $x(0) = x_{0,i}, i = 1,2,3$ depicted as (—✳—), (—▫—) and (—▲—).

Figure 2.2. Simulation results of the actuated undamped linear discrete-time oscillator (2.54) for the proposed path following rbMPC controller.

From Figure 2.2a and 2.2f, we see that all initial conditions $x(0) = x_{0,i}, i = 1,2,3$ converge to the reference path. Hence, we can conclude that the proposed rbMPC controller drives the reference path $P(\theta)$ for the closed-loop system asymptotic stable which confirms our results from Theorem 3. Moreover, we know from Theorem 4 that exact constraint satisfaction is guaranteed if the initial condition $x(0)$ satisfies $x(0) \in \hat{\mathcal{E}}_N(\delta, \theta(0))$ for some $\theta(0) \in \mathbb{R}$. Note that we calculated the set $\hat{\mathcal{E}}_N(\delta, \theta(0)$ for $\theta(0) = 0$. However, we also see that if the initial condition $x(0) = x_{0,1/3}$ does not initially lie in the set $\hat{\mathcal{E}}_N(\delta, 0)$, strict constraint satisfaction is still possible. This fact originates from the conservative bound used within the calculation of the set $\hat{\mathcal{E}}_N(\delta, 0)$. Thanks to the use of relaxed barrier functions, we can also provide initial conditions $x(0) = x_{0,3}$ that lie outside of the actual state constraint set $x(0) \notin \mathcal{X}$. Due to its global definition, the proposed rbMPC path following scheme is still able to calculate a control input such that stabilization of the reference path $P(\theta)$ for the closed-loop system can be guaranteed. Moreover, from Figure 2.2e, we can see that the rbMPC path following algorithm uses its additional degree of time reparametrization to better adapt towards the reference path. This characteristic can be seen at the time span $k \in [0, -10]$, where we see that the path parameter $\theta(k)$ is not a straight line but slightly bent within the mentioned time span.

Remark 12. We calculated the set $\hat{\mathcal{E}}_N(\delta, 0)$ by discretizing the state constraint set $\mathcal{X}$ with step size 0.01. According to Theorem 4, we checked if the current state belongs to the set $\hat{\mathcal{E}}_N(\delta, 0)$. Note that the set is only valid for initial path parameter of $\theta(0) = 0$. Note that in general, no closed-form solution to $J_\infty^*(x, \theta(0))$ for all $x \in \mathbb{R}^n$ and some $\theta(0) \in \mathbb{R}$ exists, due to its nonlinearity. We approximate the value for each x by choosing the prediction horizon $N_s \in \mathbb{N}_+$ sufficiently large which is $J_\infty^*(x, \theta(0)) \approx J_{N_s}^*(x, \theta(0))$ with $N_s >> 1$.

In this chapter, we investigated an rbMPC scheme for trajectory tracking and path following for linear discrete-time systems. We proved asymptotic stability of the reference trajectory and the reference path for the closed-loop system and derived a special set of initial conditions to ensure strict constraint satisfaction. Finally, we verified our theoretical investigation by simulation results carried out at a linear discrete-time double integrator and at a linear discrete-time actuated undamped oscillator.

3

Learning in model predictive control

The performance of the MPC schemes, see Chapter 2, depends greatly on accurate system models. In autonomous driving, there are many sources of modeling errors such as uncertainties in the cornering stiffnesses, see H. Pacejka (2012). Hence, we want to derive an MPC scheme where system models are estimated or learned online, purely based on input and output data. In the following, we will investigate model-independent MPC for linear discrete-time systems with unknown system dynamic. First, we design a model-independent receding horizon MPC scheme based on asymptotically accurate predictor maps for the closed-loop trajectory. Second, we design an estimation algorithm to estimate the closed-loop trajectory along some predefined estimation horizon. Our algorithm is related to the idea of the so-called auto regressive with exogenous input (ARX) model estimation or subspace identification technique, where in a receding horizon fashion input and signal data are used for online model identification. However, these approaches result in a least-squares solution and do not guarantee exact convergence, see I. Adelwahed, A. Mbarek and K. Bouzrara (2017). Moreover, we want to mention that we do not perform any kind of system identification but rather signal prediction of the closed-loop trajectory. Finally, we combine the model independent MPC scheme and the estimation scheme into an overall algorithm. The results presented in this chapter are based on C. Ebenbauer, F. Pfitz and S. Yu (2021).

3.1 A model-independent receding horizon control scheme

Consider the system

$$z(k+1) = Fz(k) + Gv(k), \quad z(0) = z_0,$$
$$y(k) = Hz(k) \tag{3.1}$$

with the system state $z(k) \in \mathbb{R}^n$, the control input $v(k) \in \mathbb{R}^q$, and the system output $y(k) \in \mathbb{R}^p$ at sampling instant k.

Assumption 13. We assume that **a)** (F, G) is stabilizable and (F, H) is detectable, **b)** an upper bound $m \geq n$ of the state dimension is known, and **c)** F, G, H are unknown and only output measurements are available.

In the following, we will state our model-independent controller for the control of fully unknown linear discrete-time systems in three steps. First, we will introduce a model-independent receding horizon control scheme that uses time-varying but linear and asymptotic accurate predictor maps as an internal model. Second, we design a proximity-based estimation scheme that belongs to the class of linear time-varying and asymptotic accurate predictor maps from Step 1. The results of Step 1 and Step 2 can be seen as independent. Third, we combine the results from Step 1 and Step 2 into an overall modeling-free controller for the control of fully unknown linear discrete-time systems.

A Model-independent Receding Horizon Control Scheme
Consider the system

$$x(k+1) = Ax(k) + Bu(k) \tag{3.2}$$

with the system state $x(k) \in \mathbb{R}^n$ and the control input $u(k) \in \mathbb{R}^q$ at sampling instant $k \in \mathbb{N}$. Furthermore, consider the following optimization problem

$$V_1^*(x, p_1) = \min \sum_{i=0}^{N-1} x_i^\top Q x_i + u_i^\top R u_i + \frac{\Gamma(x)}{\epsilon} x_N^\top Q_N x_N$$
$$\text{s.t. } x_i = P_i(k, x_0, u_0, ..., u_{i-1}), \, i = 1...N, \tag{3.3}$$
$$x_0 = x$$

with parameters $p_1^\top = [k, \epsilon]$, $\epsilon > 0$ along the prediction horizon $N \in \mathbb{N}_+$. The decision variables are $u_i \in \mathbb{R}^q$, $i = 0...N-1$ and $x_i \in \mathbb{R}^q$, $i = 0...N$. The map $P_i : \mathbb{N} \times \mathbb{R}^n \times \mathbb{R}^{iq} \to \mathbb{R}^n$ refers to an i-th step-ahead state (or signal) predictor. If in (3.3) we choose x to be the state of (3.2) at time instant k, i.e., $x = x(k)$ and if we choose $p_1 = p_1(k)$ at time instant k for a given sequence $\{p_1(k)\}_{k \in \mathbb{N}}$, the optimal solution is denoted by $u_i^*(x(k), p_1(k))$, $i = 0...N-1$, $x_i^*(x(k), p_1(k))$, $i = 1...N$. The closed-loop of the *receding horizon control policy* applied to (3.2) arises to

$$x(k+1) = Ax(k) + Bu_0^*(x(k), p_1(k)). \tag{3.4}$$

We impose now the following Assumptions.

Assumption 14. We assume that the prediction horizon satisifies $N \geq n = \dim(x(k))$ and $Q > 0, R > 0, Q_N > 0, \Gamma : \mathbb{R}^n \to \mathbb{R}$ positive definite.

Assumption 15. a) For any $k \in \mathbb{N}$, we assume that the state predictor maps P_i, $i = 1...N$ have the following linear structure

$$P_i(k, x_0, u_0, ..., u_{i-1}) = A_i(k)x_0 + \sum_{l=0}^{i-1} B_{i-1-l}(k)u_l, \tag{3.5}$$

where $\{A_i(k)\}_{k \in \mathbb{N}}$, $i = 0...N$ with $A_0(k) = I$, and $\{B_i(k)\}_{k \in \mathbb{N}}$, $i = 0...N - 1$, $A_i(k) \in \mathbb{R}^{n \times n}$, $B_i(k) \in \mathbb{R}^{n \times q}$, are both uniformly bounded matrix sequences such that

$$\lim_{k \to \infty} A_i(k) = \hat{A}_i, \ \lim_{k \to \infty} B_i(k) = \hat{B}_i. \tag{3.6}$$

b) Moreover, we assume that for any $x(0) \in \mathbb{R}^n$, the state predictor maps P_i along the trajectory of the closed-loop (3.3) and (3.2) with $s_l := u_l(x(k), p_1(k))$ satisfy

$$P_j(k, P_i(k, x(k), s_0, ..., s_{i-1}), s_i, .., s_{i+j-1}) = P_i(k, P_j(k, x(k), s_0, ..., s_{j-1}), s_j, .., s_{i+j-1})$$
$$= P_{j+i}(k, x(k), s_0, ..., s_{i-1}), s_i, .., s_{i+j-1}) \tag{3.7}$$

and predict asymptotically accurate with respect to system (3.2) in the sense that we have, for any $k \in \mathbb{N}$

$$A_i(k)x(k) + \sum_{l=0}^{i-1} B_{i-1-l}(k)s_l = A^i x(k) + \sum_{l=0}^{i} A^{i-1-l} B s_l + e_i(k), \ i = 0...N, \tag{3.8}$$

where the following error bounds hold for any $i = 0...N$: $\|e_i(k)\|^2 \leq \omega_1(k) + \omega_2(k)\|x(k)\|^2 + \omega_3(k) \sum_{l=0}^{i} \|s_l\|^2$ with $\lim_{k \to \infty} \omega_j(k) = 0$, $j = 1, 2, 3$. **c)** Finally, for any trajectory $x(k)$ of the closed-loop (3.4), there exists a non-negative and bounded function $\beta : \mathbb{N} \to \mathbb{R}_+$ with $\lim_{k \to \infty} \beta(k) = 0$, functions $\mu_0(k), ..., \mu_{N-1}(k)$, some $c > 0$ and γ belonging to a $\mathcal{K}_\infty$ function, such that

$$\|A_N(k)x(k) + \sum_{l=0}^{N-1} B_{N-1-l}(k)\mu_l(k)\|^2 \leq c\beta(k)$$
$$\sum_{l=0}^{N-1} \|\mu_l(k)\|^2 \leq \gamma(\|x(k)\|). \tag{3.9}$$

Remark 16. Equations (3.5) and (3.6) in Assumption 15 ensure that we may have linear time-varying but converging state predictor maps. Equation (3.7) ensures the state space property in the sense that the predictor maps commute like flow maps of (time-invariant) dynamical systems do. Equation (3.8) ensures that the predictor maps are able to accurately predict the state trajectory $x_1(x(k), p_1(k)), \ldots, x_N(x(k), p_1(k))$ in (3.3) *along the actual closed loop trajectory*. Notice that if the state predictor maps are determined (learned) online, then Assumption 15 does not imply that the knowledge of the state predictors implies a system identification in the sense that either the equation $(\hat{A}_i, \hat{B}_{l-i}) = (A^i, A^{l-i}B)$ must hold nor that for *any* initial data or *any* input sequence the predictions are (asymptotically) accurate.

The main goal of the next subsection is to show that the state of the closed-loop (3.2), (3.3) converges to zero under the stated assumptions. An important property of the proposed scheme is that the objective (and potentially constraints) can be chosen independently from the system in the sense that no terminal cost or constraint needs to be computed based on some model information or data. Only an upper bound on the state dimension needs to be known to specify the prediction horizon. This is a desirable property when controlling unknown systems, and hence we refer to the scheme as a model-independent receding horizon scheme. Furthermore, given that the system is not controllable, a zero terminal state constraint cannot be used. Instead, a novel adaptive terminal state weighting scheme is introduced, and no further stabilizing constraints/cost terms for stability are needed. Finally, the online computational burden is rather low given that the problem boils down to a regression problem (quadratic programming with equality constraints).

3.1.1 Results

We define the following auxiliary problems

$$V_2^*(x, p_2) = \min \ \xi_N^\top Q_N \xi_N$$

$$\text{s.t.} \quad \xi_{i+1} = A_{i+1}(k)x + \sum_{l=0}^{i} B_{i-l}(k)v_l, \qquad (3.10)$$

$$\xi_0 = x, \ i = 0 \ldots N - 1$$

with $p_2 = k$. A corresponding notation as for (3.3) is used in (3.10).

Theorem 5. Suppose Assumption 14 holds true. Then, we have that the solution of (3.3) is unique and parameterized in x in the sense of $x_i(x, p_1) = K_{1,i}(x, p_1)x$ and $u_i(x, p_1) = K_{3,i}(x, p_1)x$ for all $i = 0, \ldots, N$. •

Lemma 7. Consider (3.3) and suppose Assumption 15 a.) holds true. Further, suppose $\Gamma : \mathbb{R}^n \to \mathbb{R}$ is a function such that for all $x \in \mathbb{R}^n$, all $k \in \mathbb{N}$ and all $\epsilon > 0$ it holds that $\Gamma(x) \geq c(\sum_{i=0}^{N-1} \|\xi_i(x, p_2)\|^2 + \|v_i(x, p_2)\|^2)$ for some $c > 0$, where $p_2 = k$ and $\{v_i(x, p_2)\}_{i=0}^{N-1}$, $\{\xi_i(x, p_2)\}_{i=0}^{N-1}$ is some solution of (3.10). Then there exists a $\rho > 0$ such that solution $x_N^*(x, p_1)$, $p_1^\top = [k, \epsilon]$, of (3.3) satisfies for all $x \in \mathbb{R}^n$, all $k \in \mathbb{N}$, and all $\epsilon > 0$

$$x_N^*(x, p_1)^\top Q_N x_N^*(x, p_1) \leq V_2^*(x, p_2) + \epsilon \rho. \tag{3.11}$$

•

In the following, we will use the result of Lemma 12 given in Appendix A to prove Theorem 6 which basically defines an ISS property for discrete-time systems in terms of Lyapunov functions.

Theorem 6. Suppose Assumption 14 and 15 a.) - c.) hold true. Then, there exists for any initial state $x(0)$ a sequence $\{\epsilon(k)\}_{k \in \mathbb{N}}$ with $\epsilon(k) > \beta(k)$ and $\lim_{k \to \infty} \epsilon(k) = 0$ such that the state $x(k)$ of the closed loop (3.3) and (3.2) converges globally to zero, if one chooses in (3.3) $\Gamma(x(k)) = \alpha x(k)^\top x(k)$, $\alpha > 0$ and $\epsilon = \epsilon(k)$.

•

3.2 A proximity-based estimation scheme

Consider an output sequence (or some observed signal) and an input sequence

$$\{y(k)\}_{k \in \mathbb{N}}, \ \{v(k)\}_{k \in \mathbb{N}} \tag{3.12}$$

with $y(k) \in \mathbb{R}^{\bar{p}_y}$, $v(k) \in \mathbb{R}^{\bar{q}_v}$. Let

$$x(k) = \phi_y(y(k), ..., y(k - \bar{N}_y + 1)) \in \mathbb{R}^{\bar{n}},$$
$$u(k) = \phi_v(v(k), ..., v(k - \bar{N}_v + 1)) \in \mathbb{R}^{\bar{q}}, \tag{3.13}$$

and let $\phi_y : \mathbb{R}^{\bar{p}_y \bar{N}_y} \to \mathbb{R}^{\bar{n}}$, $\phi_v : \mathbb{R}^{\bar{q}_v \bar{N}_v} \to \mathbb{R}^{\bar{q}}$ be some given basis (lifting) functions, e.g., $\phi_y(y_1(k), y_2(k), y_1(k-1), y_2(k-1)) = [y_1(k), y_2(k), y_1(k-1), y_2(k-1), y_1(k)y_2(k), y_1(k)^2, y_2(k)^2]^\top$, $\bar{p} = 2$, $\bar{n} = 6$ and $\bar{N}_y = 2$. Furthermore, consider at time instant k the optimization problem

$$\theta^*(k) = \text{arg min} \ \ c(e, k) + D(\theta, \theta(k-1))$$
$$\text{s.t.} \ \ s(k) - R(k)\theta = e \tag{3.14}$$

with $\bar{N} \in \mathbb{N}$, $c : \mathbb{R}^{\tilde{n}\bar{N}} \times \mathbb{N} \to \mathbb{R}$ and $D(x,y) = g(x) - g(y) - (x-y)^{\top}\nabla_y g(y)$ defines the Bregman distance induced by a function $g : \mathbb{R}^{\tilde{n}\bar{N}} \to \mathbb{R}$. The vector $s(k)$ is defined as

$$s(k) = [x(k)^{\top} \ \dots \ x(k - \bar{N} + 1)^{\top}]^{\top}, \tag{3.15}$$

and the matrix $R(k)$ is defined as

$$R(k) = \begin{bmatrix} x(k-1)^{T} \otimes I & u(k-1)^{T} \otimes I \\ \vdots & \vdots \\ x(k-\bar{N})^{T} \otimes I & u(k-\bar{N})^{T} \otimes I \end{bmatrix}. \tag{3.16}$$

Decision variables are the parameter vector $\theta \in \mathbb{R}^{\tilde{n}(\tilde{n}+\tilde{q})}$ and e. We refer to θ when using $\arg\min$ since the (slack) variables e can be eliminated and have been introduced just for the notational convenience. Also, we define in the following $y(k) = 0, v(k) = 0, v_k = 0$ etc. whenever $k < 0$. We now impose the following assumptions.

Assumption 17. The objective function c is continuously differentiable and strictly convex in the first argument and satisfies for all k and $e \neq 0$: $c(e,k) > c(0,k)$. Furthermore, the function g, which defines the Bregman distance D, is continuously differentiable and strictly convex.

Assumption 18. For the given sequences in (3.12) and given $x(k) = \phi_y(y(k), \dots, y(k - \bar{N}_y + 1))$ and $u(k) = \phi_v(v(k), \dots, v(k - \bar{N}_v + 1))$, there exist matrices A, B $x_0 \in \mathbb{R}^{\tilde{n}}$ and such that satisfy

$$x(k+1) = Ax(k) + Bu(k), \quad x(0) = x_0. \tag{3.17}$$

Remark 19. a) Notice that $s(k) = R(k)\theta$ with $\theta^{\top} = [\text{vec}(A)^{\top}, \text{vec}(B)^{\top}]$, where $\text{vec}(A)$ corresponds to the (column-wise) vectorization of a matrix A, is the linear system of equations $x(j) = Ax(j-1) + Bu(j-1)$, $j = k \dots k - \bar{N} + 1$. b) If $c(e,k) = \|e\|^2, g(x) = \|x\|^2$, then (3.16) reduces to a least squares parameter estimation problem, where a closed form solution to it is known. The motivation for a general convex cost is its flexibility in tuning the estimator. Similarly as in a recently proposed state estimation scheme based on proximal minimization M. Gharbi, B. Gharesifard and C. Ebenbauer (2020), specifying different c, D allows taking into account various aspects like outliers in the data, sparsity in the parameters, or cost-biased objectives (S. Bittanti and M. Campi (2006)).

Remark 20. Assumption 18 imposes that the given (lifted) signal $\{x(k)\}_{k \in \mathbb{N}}$ can be reproduced by some LTI system with the given (lifted) input sequence

$\{u(k)\}_{k\in\mathbb{N}}$. Notice that reproducing a given signal by (3.17) does not imply that the signal $x(k)$ itself originates from (3.17) nor by an LTI system at all. For example, a given *(single)* trajectory of a nonlinear system, or even all trajectories of a large class of nonlinear systems, can be reproduced by or embedded into high dimensional linear (not necessarily controllable) systems using, for example, Carleman or Koopman lifting techniques. Hence, (3.17) represents a signal model (of the actual trajectory) rather than a system model (of all possible trajectories). This means that different actual trajectories can coincide with different signal models but with the same system model.

The main goal of the next subsection is to show that the (parameter) estimates $\theta^*(k)$ obtained from (3.14) converge and that the estimates can be used to define ith step-ahead signal predictor maps $v_i = P_i(k, v_0, u_0, ..., u_{i-1})$, which have the properties as described in Assumption 15 for the signal model (3.17) and for the given data (3.13). In summary, we therefore aim for an estimation scheme to obtain asymptotically accurate predictor maps for a given input and output sequence that can be embedded in a potentially high-dimensional linear signal model. An important property of the proposed scheme is that no system identification is carried out and no persistency of excitation condition is needed. The online computational burden is again rather low since, in its simplest form, the problem boils down to a least-squares regression problem.

3.2.1 Results

We first prove a lemma that is a key step for the convergence of the estimation scheme. It provides a convergence result for a proximal minimization scheme with a time-varying objective function that has at least one common (time-invariant) minimizer.

Lemma 8. Let $f : \mathbb{R}^n \times \mathbb{N} \to \mathbb{R}$ be convex and continuous differentiable in the first argument. Suppose the set of common minimizers $\mathcal{X} = \{x^* \in \mathbb{R}^n : f(x, k) \geq f(x^*, k) := 0 \; \forall x \in \mathbb{R}^n\}$ of f is non-empty. Let further $g : \mathbb{R}^n \to \mathbb{R}$ be strictly convex, continuous differentiable, and let D denote the Bregman distance induced by g. Assume that D is convex in the second argument, then the proximal minimization iterations x_k^* given by

$$x_{k+1}^* = \arg\min_x f(x,k) + D(x, x_k^*) \tag{3.18}$$

converge to a minimizer $x^* \in \mathbb{R}^n$ such that $\lim_{k\to\infty} f(x_k^*, k) = f(x^*, k) = 0$. •

Notice that the Bregman distance is in general not convex in the second argument, but there are important cases, such as $g(x) = x^T Q x$, $Q > 0$, where this holds (H. Bauschke and J. Borwein (2001)). The next theorem is the main result of this subsection.

Theorem 7. Consider the sequences (3.12) with some given basis functions (3.13) and consider the optimization problem (3.14). Suppose Assumptions 17 and 18 hold true and assume that D is convex in the second argument. Then the following statements hold true.

(i) The solution sequence $\{\theta^*(k)\}_{k \in \mathbb{N}}$ converges, i.e. $\lim_{k \to \infty} \theta^*(k) = \theta^*$ and is uniformly bounded.

(ii) If one defines $\theta^*(k)^\top = [\text{vec}(A(k))^\top, \text{vec}(B(k))^\top]$ and predictor maps (3.5) according to

$$
\begin{aligned}
A_i(k) &= A(k)^i \\
B_i(k) &= A(k)^i B(k),
\end{aligned}
\tag{3.19}
$$

$i = 0...\bar{N}$, then the predictor fulfills the properties (3.6), (3.7) and (3.8) in Assumption 15 a.) - b.) with respect to the signal model (3.17) and the sequences (3.13).

$\bullet$

Remark 21. Assumption 15 c.) is the main technical assumption imposed in the proposed scheme. Notice that 7 (i) and (ii) proves that the predictor fulfills Assumption 15 a.) - b.). Additionally, Assumption 15 c.) is needed to ensure that for any initial state $z(0)$, the state $z(k)$ of the closed loop for the overall scheme described in Theorem 8, converges to zero as k goes to infinity. We can think of the first eq. in Assumption 15 c.) as an asymptotic stabilizability property of the predicted matrix sequence $\{A_i(k)\}_{k \in \mathbb{N}}, \{B_i(k)\}_{k \in \mathbb{N}}$. Moreover, the second eq. of Assumption 15 c.) refers to some stabilizing control law with finite energy which is dependent on the system state. Note that if the limit system $\hat{A}_i, \hat{B}_i$ is stabilizable and if there exists some non-empty set (which can be arbitrary small) around the limit system in which the stabilizability condition is preserved, then there exists some $\bar{k} \in \mathbb{N}$ such that for $k \geq \bar{k}$, it holds that Assumption 15 c.) is satisfied.

3.3 Overall scheme

Lemma 9. Consider an arbitrary output and input sequence $\{y(k)\}_{k \in \mathbb{N}}, \{v(k)\}_{k \in \mathbb{N}}$ of system (3.1), and suppose Assumption 13 holds true. Let

$$
\begin{aligned}
x(k) &= \phi_y(y(k), ..., y(k - m + 1)) = \left[y(k)^\top, ..., y(k - m + 1)^\top \right]^\top, \\
u(k) &= \phi_v(v(k), ..., v(k - m + 1)) = \left[v(k)^\top, ..., v(k - m + 1)^\top \right]^\top.
\end{aligned}
\tag{3.20}
$$

Then the sequences $\{u(k)\}_{k \in \mathbb{N}}$, $\{x(k)\}_{k \in \mathbb{N}}$ satisfy Assumption 18.   •

Remark 22. Suppose the initial condition is chosen to $x(0) = [z(0)^\top, 0^\top, \ldots, 0^\top]^\top$. Moreover, suppose that the output and input sequences $\{y(k)\}_{k \in \mathbb{N}}$, $\{v(k)\}_{k \in \mathbb{N}}$ are defined according to eq. (3.13). If we choose $v(k) = K_s z_s(k)$ such that $F_s + G_s K_s$ is stable, which is possible since (F_s, G_s) is stabilizable, we know that that $\lim_{k \to \infty} z(k) = 0$. Thus, we can conclude that $\{x(k)\}_{k \in \mathbb{N}}$, $\{u(k)\}_{k \in \mathbb{N}}$ converges to zero as k goes to infinity as the input vector $u(k)$ and output vector $y(k)$ contain past values of the actual input $v(k)$ (modeled as an integrator chain) and past values of the actual output $y(k)$, respectively. Hence, we know that the lifted system (3.17) is stabilizable in the sense that there exists an input sequence with $v(k) = K_s z_s(k)$ such that the state $x(k)$ of the closed loop converges to zero.

Remark 23. Notice that the lifted input vector $u(k)$ in (3.20) contains past values of the actual input vector $v(k)$. To obtain a state-space model with input $v(k)$, one can simply add state variables to the signal model. In more detail, define an integrator chain dynamics of the form $\zeta_1(k+1) = \zeta_2(k), \ldots, \zeta_{m-2}(k+1) = \zeta_m(k), \zeta_{m-1}(k+1) = v(k)$, hence $\zeta_1(k)$ corresponds to $v(k-m)$ and so on. This augmentation does not affect the stabilizability property, since the states of the integrator chain converge to zero if $v(k)$ converges to zero. This state augmentation in the signal model leads to matrices with at least the size $A \in \mathbb{R}^{mp+(m-1)q \times mp+(m-1)q}$, $B \in \mathbb{R}^{mp+(m-1)q \times q}$ and needs to be taken into account when implementing the receding horizon scheme.

We are now ready to close the loop. Based on Lemma 9, we know that the output and input sequences $\{y(k)\}_{k \in \mathbb{N}}$, $\{v(k)\}_{k \in \mathbb{N}}$ of system (3.1) satisfy Assumption 18 with regard to the signal model (3.17), see Lemma 9. Assumption 14 and Assumption 17 can be satisfied by setting up the optimization problem accordingly. By Theorem 7, the estimation scheme (3.14) guarantees that Assumptions 15 a.) - b.) is satisfied. Thus the receding horizon scheme guarantees, by Theorem 6 with $\Gamma(x) = \alpha x^T x, \alpha > 0$ and a sequence $\epsilon(k) > \beta(k)$ for any $k \in \mathbb{N}$ according to Assumption 15 c.) and with Lemma 9, that the state and the input of (3.1) converge to zero. These arguments lead to the next theorem.

Theorem 8. Consider the closed-loop system consisting of the system (3.1), the receding horizon control scheme (3.3), and the proximity-based estimation scheme (3.14). Define $x(k)$, $u(k)$ (ϕ_y, ϕ_v) according to equation (3.20) and set up the predictor scheme (3.14) according to Assumption 17. Further, set up the receding horizon scheme (3.3) according to Assumption 14 with $n = m$, $\alpha > 0, \Gamma(x) = \alpha x^T x$. Then, under Assumption 13 and 15 c.), for any initial state

$z(0)$, there exists a sequence $\epsilon(k)$ with $\{\epsilon(k)\}_{k\in\mathbb{N}}$, $\epsilon(k) > \beta(k)$ (according to Assumption 15 c.)) with $\lim_{k\to\infty} \epsilon(k) = 0$, such that the state $z(k)$ of the closed loop and the input $v(k)$ of the closed loop converges to zero as k goes to infinity.
●

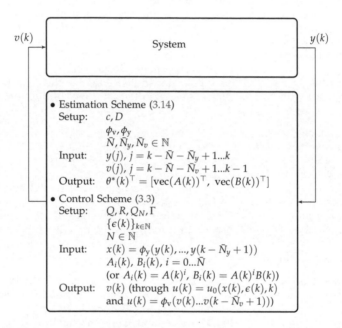

Figure 3.1. The overall scheme for stabilization of the origin for LTI systems with unknown system dynamics.

We conclude with some remarks. First, notice that the state extension in Remark 23 has to be taken into account when implementing the overall scheme. Moreover, if ϕ_v, ϕ_y are nonlinear functions (notice that in (3.20) these are linear functions), then this requires further considerations (e.g., one has to be able to extract the input v from ϕ_v), which is outside of the scope of this work. Second, notice that the convergence result of Theorem 8 also holds for certain classes of nonlinear systems. As already mentioned in Remark 20, certain (trajectories of) nonlinear systems can be embedded in high-dimensional linear (signal) models, such as $z_1(k+1) = z_1(k) + z_2(k)^2 + u(k), z_2(k+1) = 0.5z_2(k)$, which can be written as a linear system with an additional state variable $z_3(k) = z_2(k)^2$, see B. Koopman (1931); C. Münzing (2017); M. Korda and I. Mezic (2017); T. Carleman (1932). However, a finite linear representation of nonlinear system via the Koopman operator is only possible if there is a single isolated fixed point, see S. Brunton et al. (2016). Third, if a model of (3.1) is known, then it follows from the proofs of Theorems 8 that global ISS property of the closed loop at the origin can be guaranteed. Moreover, in applications, models are often available for at least some parts of the system. The proposed scheme can eventually be applied to such situations, for example, if we have two subsystems $x_1(k+1) = F_1 x_1(k) + F_2 x_2(k) + G_1 u(k), x_2(k+1) = F_3 x_1(k) + F_4 x_2(k) + G_2 u(k), y(k) = x_1(k)$, where F_3, F_4, G_2 is unknown, than one can consider the second subsystem as the unknown system with input $u(k), x_1(k)$ and adjust the receding horizon control scheme accordingly. It is also possible to model unknown disturbances or reference signals by a so-called unknown exosystem and to treat this system as a signal model, see C. Münzing (2017). Finally, a priori knowledge about the system model can be taken into account by initializing the estimator appropriately or by including parameter constraints in the estimator.

3.4 Numerical example

In this section, we show simulation results of the overall scheme for different dynamic systems. First, we choose a *linear stabilizable system* which shows the ability to deal with stabilizable systems and unstable modes. More precisely, we consider

$$
z(k+1) = \begin{bmatrix} 1.01 & 0 \\ 0 & 0.92 \end{bmatrix} z(k) + \begin{bmatrix} -0.1 \\ 0 \end{bmatrix} v(k), \qquad z(0) = \begin{bmatrix} 5 & -0.1 \end{bmatrix}^\top,
$$
$$
y(k) = \begin{bmatrix} 1 & 0 \\ 0 & 1 \end{bmatrix} z(k),
$$

(3.21)

where $y(k) \in \mathbb{R}^2$, $z(k) \in \mathbb{R}^2$, and $v(k) \in \mathbb{R}$ refers to the system output, the system state, and the control input at sampling instant $k \in \mathbb{N}$, respectively. Hence, $\bar{p}_y = 2$ and $\bar{q}_v = 1$. Furthermore, we do not impose any state or input constraints. The parameters of the estimation and the MPC algorithm are given in Table 3.1.

Estimation algorithm	Control algorithm
$c(x) = \|x\|^2$	$Q = 100 I_{\bar{N}_y \bar{p}_y \times \bar{N}_y \bar{p}_y}$
$g(x) = \|x\|^2$	$R = 10 I_{\bar{N}_v \bar{q}_v \times \bar{N}_v \bar{q}_v}$
$\Phi_v = [v(k)^\top, ..., v(k - \bar{N}_v + 1)^\top]^\top$	$Q_N = 0.001 I_{\bar{N}_y \bar{p}_y \times \bar{N}_y \bar{p}_y}$
$\Phi_y = [y(k)^\top, ..., y(k - \bar{N}_y + 1)^\top]^\top$	$\Gamma(x) = x^\top x$
$\bar{N} = 10$	$N = 3$
$(\bar{N}_y, \bar{N}_v) = (1, 1)$	$\epsilon(k) = \frac{1}{1 + 10k}$

Table 3.1. Tuning variables for the estimation and the MPC algorithm. For the estimation algorithm, we choose the cost term c, the function of the Bregman distance g, the input basis function Φ_v, the output basis function Φ_y, the estimation horizon $\bar{N}$, the horizon length of the output signal $\bar{N}_y$ and the horizon length of the input signal $\bar{N}_v$. For the MPC algorithm, we choose the state weight matrix Q, the input weight matrix R, the terminal state weight matrix Q_N, the terminal state weighting Γ, the prediction horizon N, and the sequence $\epsilon(k)$.

In Figure 3.2, we show simulation results of the combined algorithm depicted in (3.1) for the unknown system (3.21).

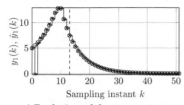

a) Evolution of the true system output $y_1(k)$ (–⊙–) and the estimated system output $\hat{y}_1(k)$ (–∗–).

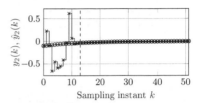

b) Evolution of the true system output $y_2(k)$ (–⊙–) and the estimated system output $\hat{y}_2(k)$ (–∗–).

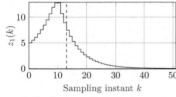

c) Evolution of the system state $z_1(k)$.

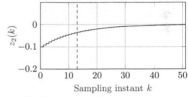

d) Evolution of the system state $z_2(k)$.

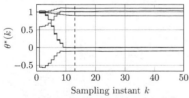

e) Evolution of the estimation vector $\theta^*(k)$.

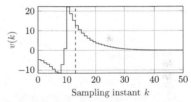

f) Evolution of the control input $v(k)$.

Figure 3.2. Results of the overall algorithm depicted in Figure 3.1 for the control of a fully unknown linear stabilizable discrete-time system defined in (3.21). The figure depicts the evolution of the true system output $y(k)$, the estimated system output $\hat{y}(k)$, the evolution of the unknown system state $z(k)$, and the evolution of the estimation vector $\theta^*(k)$. We see that after an initial learning phase, the estimation vector converges to a constant vector and the system state and the system output converge to zero.

Next, we choose a *linear stabilizable system* where the output does not contain every state. Hence, we choose to system to be

$$
z(k+1) = \begin{bmatrix} 0 & 1 & 0.1 \\ 0 & 1.02 & 0 \\ 0 & 0 & 0.92 \end{bmatrix} z(k) + \begin{bmatrix} 0 \\ 1 \\ 0 \end{bmatrix} v(k), \qquad z(0) = \begin{bmatrix} 0.1 & 0.1 & -10 \end{bmatrix}^\top,
$$
$$
y(k) = \begin{bmatrix} 1 & 0 & 1 \end{bmatrix} z(k),
$$

$$(3.22)$$

where $y(k) \in \mathbb{R}$, $z(k) \in \mathbb{R}^3$, and $v(k) \in \mathbb{R}$ refer to the system output, the system state, and the control input at sampling instant k, respectively. Hence, we have $\bar{p}_y = 1$ and $\bar{q}_v = 1$. The parameters of estimation and MPC algorithm are given in Table 3.2.

Estimation algorithm	Control algorithm
$c(x) = \|\|x\|\|^2$	$Q = 100 I_{\bar{N}_y \bar{p}_y \times \bar{N}_y \bar{p}_y}$
$g(x) = \|\|x\|\|^2$	$R = 10000 I_{\bar{N}_v \bar{q}_v \times \bar{N}_v \bar{q}_v}$
$\Phi_v = [v(k)^\top, ..., v(k - \bar{N}_v + 1)^\top]^\top$	$Q_N = 100 I_{\bar{N}_y \bar{p}_y \times \bar{N}_y \bar{p}_y}$
$\Phi_y = [y(k)^\top, ..., y(k - \bar{N}_y + 1)^\top]^\top$	$\Gamma(x) = x^\top x$
$\bar{N} = 8$	$N = 20$
$(\bar{N}_y, \bar{N}_v) = (4, 4)$	$\epsilon(k) = \frac{1}{1 + 1000k}$

Table 3.2. Tuning variables for the estimation and the MPC algorithm. For the estimation algorithm, we choose the cost term c, the function of the Bregman distance g, the input basis function Φ_v, the output basis function Φ_y, the estimation horizon $\bar{N}$, the horizon length of the output signal $\bar{N}_y$ and the horizon length of the input signal $\bar{N}_v$. For the MPC algorithm, we choose the state weight matrix Q, the input weight matrix R, the terminal state weight matrix Q_N, the terminal state weighting Γ, the prediction horizon N, and the sequence $\epsilon(k)$.

In Figure 3.3, we show simulation results of the combined algorithm depicted in (3.1) for the unknown system (3.22).

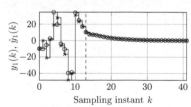

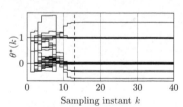

a) Evolution of the true system output $y_1(k)$ (–⊖–) and the estimated system output $\hat{y}_1(k)$ (–✳–).

b) Evolution of the estimation vector $\theta^*(k)$.

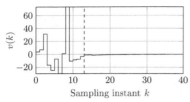

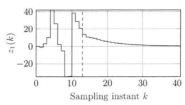

c) Evolution of the control input $v(k)$.

d) Evolution of the system state $z_1(k)$.

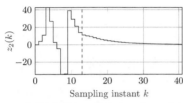

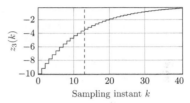

e) Evolution of the system state $z_2(k)$.

f) Evolution of the system state $z_3(k)$.

Figure 3.3. Results of the overall algorithm depicted in Figure 3.1 for the control of the fully unknown system defined in (3.22). The figure depicts the evolution of the true system output $y(k)$, the estimated system output $\hat{y}(k)$, the evolution of the unknown system state $z(k)$, and the evolution of the estimation vector $\theta^*(k)$. We see like in the previous example that after an initial learning phase the estimation vector converges to a constant vector and the system state and system output converge to zero.

Next, we want to control a system that is originating from a nonlinear system. We choose the nonlinear system to be the *Van-der-Pol oscillator*. The continuous-time nonlinear state equations are denoted as

$$\begin{bmatrix} \dot{z}_1 \\ \dot{z}_2 \end{bmatrix} = \begin{bmatrix} z_2 \\ \epsilon(1 - z_1^2)z_2 - z_1 + v \end{bmatrix}, \quad z(0) = \begin{bmatrix} -1 & 1.7 \end{bmatrix}^\top, \tag{3.23}$$
$$y = z,$$

where $y \in \mathbb{R}^2$, $z \in \mathbb{R}^2$ and $v \in \mathbb{R}$ refer to the system output, the system state, and the control input. Hence, we have that $\bar{p}_y = 2$ and $\bar{q}_v = 1$. Based on the well-known theory about the Van-der-Pol oscillator, we know that for $\epsilon > 0$ and $u = 0$, the set point $(z_s, u_s) = (0, 0)$ is unstable. To arrive at a nonlinear discretized system, we Euler-forward discretize (3.23) with a step size of $h = 0.1$. Again, we do not impose any state and input constraints. The parameters of the estimation and MPC algorithm are given in Table 3.3.

Estimation algorithm	Control algorithm
$c(x) = \|x\|^2$	$Q = 100 I_{\bar{N}_y \bar{p}_y \times \bar{N}_y \bar{p}_y}$
$g(x) = \|x\|^2$	$R = 100 I_{\bar{N}_v \bar{q}_v \times \bar{N}_v \bar{q}_v}$
$\Phi_v = [v(k)^\top, ..., v(k - \bar{N}_v + 1)^\top]^\top$	$Q_N = 1 I_{\bar{N}_y \bar{p}_y \times \bar{N}_y \bar{p}_y}$
$\Phi_y = [y(k)^\top, ..., y(k - \bar{N}_y + 1)^\top]^\top$	$\Gamma(x) = x^\top x$
$\bar{N} = 10$	$N = 10$
$(\bar{N}_y, \bar{N}_v) = (2, 2)$	$\epsilon(k) = \frac{1}{1+k}$

Table 3.3. Tuning variables for the estimation and the MPC algorithm. For the estimation algorithm, we choose the cost term c, the function of the Bregman distance g, the input basis function Φ_v, the output basis function Φ_y, the estimation horizon $\bar{N}$, the horizon length of the output signal $\bar{N}_y$ and the horizon length of the input signal $\bar{N}_v$. For the MPC algorithm, we choose the state weight matrix Q, the input weight matrix R, the terminal state weight matrix Q_N, the terminal state weighting Γ, the prediction horizon N, and the sequence $\epsilon(k)$.

In Figure 3.4, we show simulation results of the combined algorithm depicted in (3.1) for the Euler forward discretized unknown system (3.23).

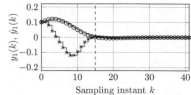

a) Evolution of the true system output $y_1(k)$ (–⊖–) and the estimated system output $\hat{y}_1(k)$ (–✳–).

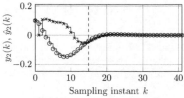

b) Evolution of the true system output $y_2(k)$ (–⊖–) and the estimated system output $\hat{y}_2(k)$ (–✳–).

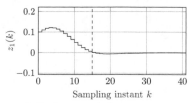

c) Evolution of the system state $z_1(k)$.

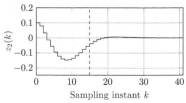

d) Evolution of the system state $z_2(k)$.

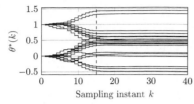
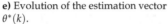
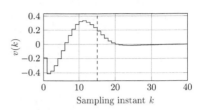

e) Evolution of the estimation vector $\theta^*(k)$.

f) Evolution of the control input $v(k)$.

Figure 3.4. Results of the overall algorithm depicted in Figure 3.1 for a fully uncertain *Van-der-Pol oscillator*. Depicted are the evolution of the true system output $y(k)$, the estimated system output $\hat{y}(k)$, the evolution of the unknown system state $z(k)$, and the evolution of the estimation vector $\theta^*(k)$. We see that after an initial learning phase, the estimation vector converges to a constant vector and the system state and the system output converge to zero.

In the following, we want to show the performance of the proposed algorithm at the physical example of a *single-link robot with a DC motor*. Consider the continuous-time single-link robot with DC motor proposed in C. Aguilar-Avelar and J. Moreno-Valenzuela (2017). More precisely, we have

$$\begin{bmatrix} \dot{z}_1 \\ \dot{z}_2 \\ \dot{z}_3 \end{bmatrix} = \begin{bmatrix} z_2 \\ -1.7382z_2 - 1309\sin(z_1) + 36.3636z_3 \\ -1000z_3 - 3.64z_2 + 100v \end{bmatrix}, \qquad z(0) = [5, -5, 1]^\top,$$
$$y = z,$$

$$(3.24)$$

where $y \in \mathbb{R}^3$, $z \in \mathbb{R}^3$, and $v \in \mathbb{R}$ refer to the system output, the system state, and the control input, respectively. Again, we Euler-forward discretize system (3.24) with step size $h = 0.01$. For the sake of clarity, the parameters of estimation and MPC algorithm are given in Table 3.4.

Estimation algorithm	Control algorithm
$c(x) = \|x\|^2$	$Q = 10I_{\bar{N}_y \bar{p}_y \times \bar{N}_y \bar{p}_y}$
$g(x) = \|x\|^2$	$R = 100I_{\bar{N}_v \bar{q}_v \times \bar{N}_v \bar{q}_v}$
$\Phi_v = [v(k)^\top, ..., v(k - \bar{N}_v + 1)^\top]^\top$	$Q_N = 10I_{\bar{N}_y \bar{p}_y \times \bar{N}_y \bar{p}_y}$
$\Phi_y = [y(k)^\top, ..., y(k - \bar{N}_y + 1)^\top]^\top$	$\Gamma(x) = x^\top x$
$\bar{N} = 10$	$N = 15$
$(\bar{N}_y, \bar{N}_v) = (2, 2)$	$\epsilon(k) = \frac{1}{10+k}$

Table 3.4. Tuning variables for the estimation and the MPC algorithm. For the estimation algorithm, we choose the cost term c, the function of the Bregman distance g, the input basis function Φ_v, the output basis function Φ_y, the estimation horizon $\bar{N}$, the horizon length of the output signal $\bar{N}_y$ and the horizon length of the input signal $\bar{N}_v$. For the MPC algorithm, we choose the state weight matrix Q, the input weight matrix R, the terminal state weight matrix Q_N, the terminal state weighting Γ, the prediction horizon N, and the sequence $\epsilon(k)$.

In Figure 3.5, we show simulation results of the combined algorithm depicted in (3.1) for the Euler forward discretized unknown system (3.24).

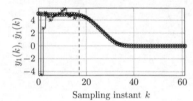

a) Evolution of the true system output $y_1(k)$ (–○–) and the estimated system output $\hat{y}_1(k)$ (–✳–).

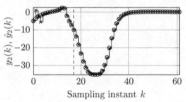

b) Evolution of the true system output $y_2(k)$ (–○–) and the estimated system output $\hat{y}_2(k)$ (–✳–).

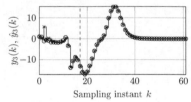

c) Evolution of the true system output $y_3(k)$ (–○–) and the estimated system output $\hat{y}_3(k)$ (–✳–).

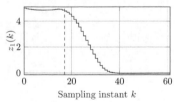

d) Evolution of the system state $z_1(k)$.

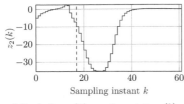

e) Evolution of the system state $z_2(k)$.

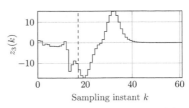

f) Evolution of the system state $z_3(k)$.

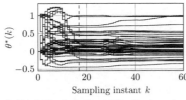

g) Evolution of the estimation vector $\theta^*(k)$.

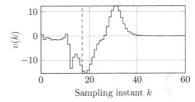

h) Evolution of the control input $v(k)$.

Figure 3.5. Results of the overall algorithm depicted in Figure 3.1 for a fully uncertain *single-link robot with DC motor*. Depicted are the evolution of the true system output $y(k)$, the estimated system output $\hat{y}(k)$, the evolution of the unknown system state $z(k)$, and the evolution of the estimation vector $\theta^*(k)$. We see that after an initial learning phase, the estimation vector converges to a constant vector and the system state and the system output converge to zero.

In this chapter, we investigated the control of fully unknown linear discrete-time systems. We divided the overall problem into an estimation problem and a control problem. In terms of the estimation problem, we did do not perform any kind of system identification. The estimator can rather be seen as a signal predictor of the closed-loop trajectory, see Remark 20. Our developed model-independent control algorithm can be classified as receding horizon control algorithm. The system model is modeled as a state predictor map which is defined as a linear time-varying system that is converging to the signal model of the closed-loop trajectory, see Assumption 15. Finally, we show that the estimates of the estimation algorithm fulfill certain conditions such that the estimator algorithm and the control algorithm can be combined into an overall algorithm which is shown in Figure 3.1. We showed convergence to the origin for fully unknown linear discrete-time systems, see Theorem 8. Our main technical assumption is Assumption 15 c.) which is fulfilled if the estimated limit system is stabilizable, see Remark 21. Finally, we apply the overall algorithm to different linear and nonlinear systems in simulation. In each example, we see that after an initial learning phase, the overall algorithm is able to steer the system output to the origin for the closed-loop system. We like to mention that our convergence guarantees are only valid for linear discrete-time system.

4

Application to automated driving

In this chapter, we investigate different dynamical models that belong to the class of single track models. We highlight its importance and adapt the rbMPC schemes proposed in Chapter 2 for the problem of an automated driving vehicle. Furthermore, based on simulations, we compare the proposed rbMPC schemes with a benchmark controller with respect to different performance criteria for different maneuvers. Finally, we implement and test the best performing algorithm at a real vehicle.

4.1 Kinematic and kinetic single track model

In the following, we will derive two dynamical models which will be either used as a system model for the rbMPC schemes, see Chapter 2, or will serve as a continuous-time simulation model. We begin this section discussing about the choice of a system model such that the rbMPC schemes achieve a satisfying performance. Basically, a good choice for a system model is a model that is complex enough, describing the crucial system behavior, and at the same time is simple enough such that the computational effort stays affordable. In literature, the class of *single track* models fulfills the aforementioned characteristic, see C. Feller (2017); J. Kong et al. (2015); M. Althoff (2018); P. Falcone et al. (2007). To achieve computational affordability, the class of single track models has certain simplifications, it neglects the roll and pitch dynamics of the vehicle because it is modeled as a rigid body, assumed to be entirely flat. Moreover, the front and rear wheel pairs are lumped into one wheel, which also explains the term single track model, see R. Rajamani (2012). Additionally, the class of single track models can be further divided into the class of kinematic and kinetic single track models. Compared to the kinetic single track model, the kinematic single track model does not take any tire slip at the front and rear wheel into account, which represents its major simplification. Thus, neglecting the acting forces and torques, the kinematic single track model can be parametrized purely depending on the geometry of the vehicle. Based on the evaluations made in J. Kong et al. (2015), we choose the kinematic single track model as a system model for the rbMPC schemes. Further, we use the kinetic single track model as a continuous-time simulation model. Next, we will introduce the reader to the discretized nonlinear differential equations of the kinematic and kinetic single track model. Additionally, we provide differential equations of the kinematic single track model that are linearized around some arbitrary reference system state and reference control input.

4.1.1 Kinematic single track model

A graphic illustration of the kinematic single track model is given in Figure 4.1. The position in the inertial frame (X, Y) is denoted as (x, y), the inertial yaw angle of the rigid body is ψ, and β refers to the heading angle of the velocity v, also called side slip angle. It is assumed that we have no rear-wheel steering $\delta_r = 0$. Thus, the rear tire points in the direction of the vehicle's body. Owing to neglect the tire slip angles, the rear and front velocities v_r and v_f point along the rear and front tires. Furthermore, every planar movement of a body can be instantly described as a pure rotation around the instantaneous center of rotation (ICR), which is also depicted in Figure 4.1, together with the rear wheel turning radius r_r.

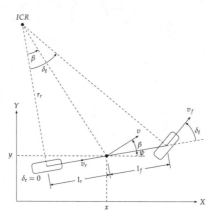

Figure 4.1. Kinematic single track model at position (x, y) with yaw angle ψ, sideslip angle β, rear and front velocity v_r, v_f, velocity at the center of mass v, front and rear steering angles $\delta_r = 0$ and δ_f, ICR and the rear wheel turning radius r_r.

Based on trigonometry, we can conclude that $r_r = \frac{\tan(\delta_f)}{l_f + l_r}$ and $\beta = \arctan(\frac{l_r}{r_r})$. Combining both equations yield to $\beta = \arctan(\frac{l_r}{l_f + l_r}\tan(\delta_f))$. The overall continuous-time nonlinear dynamics of the kinematic single track model are defined as

$$
\begin{aligned}
\dot{x} &= v\cos(\psi + \beta), \\
\dot{y} &= v\sin(\psi + \beta), \\
\dot{\psi} &= \frac{v}{l_r}\sin(\beta), \\
\dot{v} &= a, \\
\beta &= \arctan(\frac{l_r}{l_f + l_r}\tan(\delta_f)),
\end{aligned}
\tag{4.1}
$$

with the system state $\xi = [x\ y\ \psi\ v]^\top \in \mathbb{R}^4$ and the control input $u = [\beta\ a]^\top \in \mathbb{R}^2$. Note that the side slip angle β serves as a virtual control input. Moreover, note that the parameters that are needed to parametrize the kinematic single track model are the distances from the front and rear axle to the center of mass l_f and l_r, respectively. A precise derivation of the kinematic bycicle model can be found in R. Rajamani (2012). Next, we want to use the derived nonlinear kinematic single track model as a system model for the discrete-time rbMPC schemes. Therefore, we first discretize the derived continuous-time nonlinear model (4.1) Euler-forward with sampling time $h \in \mathbb{R}_{++}$, see also P. Falcone et al. (2008) and C. Feller (2017). Thus, we get $\xi(k+1) = f(\xi(k), u(k))$ with

$$
f(\xi(k), u(k)) = \begin{bmatrix}
x(k) + h\,v(k)\cos(\psi(k) + \beta(k)) \\
y(k) + h\,v(k)\sin(\psi(k) + \beta(k)) \\
\psi(k) + h\,\frac{v(k)}{l_r}\sin(\beta(k)) \\
v(k) + h\,a(k)
\end{bmatrix},
\tag{4.2}
$$

where the system state and the control input are denoted as $\xi(k) = [x(k)\ y(k)\ \psi(k)\ v(k)]^\top \in \mathbb{R}^4$ and $u(k) = [\beta(k)\ a(k)]^\top \in \mathbb{R}^2$. Second, we linearize eq. (4.2) around some arbitrary reference state $\xi_{\text{ref}}(k) = [x_{\text{ref}}(k)\ y_{\text{ref}}(k)\ \psi_{\text{ref}}(k)\ v_{\text{ref}}(k)]^\top \in \mathbb{R}^4$ and reference input $u_{\text{ref}}(k) = [\beta_{\text{ref}}(k)\ a_{\text{ref}}(k)]^\top \in \mathbb{R}^2$. The linearized system dynamics around $(\xi_{\text{ref}}(k), u_{\text{ref}}(k))$ are given as $\Delta\xi(k+1) = A(k)\Delta\xi(k) +$

$B(k)\Delta u(k)$ with

$$A(k) = \begin{bmatrix} 1 & 0 & -h\, v_{\mathrm{ref}}(k)\sin(\gamma_{\mathrm{ref}}(k)) & h\cos(\gamma_{\mathrm{ref}}(k)) \\ 0 & 1 & h\, v_{\mathrm{ref}}(k)\cos(\gamma_{\mathrm{ref}}(k)) & h\sin(\gamma_{\mathrm{ref}}(k)) \\ 0 & 0 & 1 & \frac{h}{l_r}\sin(\beta_{\mathrm{ref}}(k)) \\ 0 & 0 & 0 & 1 \end{bmatrix}, \tag{4.3}$$

$$B(k) = \begin{bmatrix} 0 & -h\, v_{\mathrm{ref}}(k)\sin(\psi_{\mathrm{ref}}(k)+\beta_{\mathrm{ref}}(k)) \\ 0 & h\, v_{\mathrm{ref}}(k)\cos(\psi_{\mathrm{ref}}(k)+\beta_{\mathrm{ref}}(k)) \\ 0 & h\,\frac{v_{\mathrm{ref}}(k)}{l_r}\cos(\beta_{\mathrm{ref}}(k)) \\ h & 0 \end{bmatrix}, \tag{4.4}$$

where $\gamma_{\mathrm{ref}}(k) = \psi_{\mathrm{ref}}(k) + \beta_{\mathrm{ref}}(k)$, $\Delta\xi(k) = \xi(k) - \xi_{\mathrm{ref}}(k)$ and $\Delta u(k) = u(k) - u_{\mathrm{ref}}(k)$ for any $k \in \mathbb{N}$.

4.1.2 Kinetic single track model

Given that the kinematic single track model (4.2) neglects the tire slips at the front and rear tire, important driving dynamic effects such as oversteering and understeering are not considered. Hence, the realistic behavior of the simulation model for higher lateral accelerations is not guaranteed. Thus, we will introduce the kinetic single track model which explicitly takes the tire slip into account. According to M. Mitschke and H. Wallentowitz (1972)[p. 713], the kinetic single track model with linear tire characteristics is only valid for lateral accelerations up to 0.4g. Because we want to be able to test our algorithms up to the limits of vehicle dynamics, we propose here to use nonlinear tire characteristics according to the Pacejka formulation, see H. Pacejka (2012). A graphical illustration of the kinetic single track model is given in Figure 4.2.

The position in the inertial frame (X, Y) is denoted as (x, y), the inertial yaw angle of the rigid body is ψ, and β refers to the heading angle of the velocity v from the vehicle longitudinal axis, the so-called side slip angle. Moreover, it is assumed that we have no rear wheel steering $\delta_r = 0$. In contrast to the kinematic single track, slip angles at the front and rear tire α_r and α_f occur. Thus, the rear and front velocities v_r and v_f do not point into the heading direction of the respective rear and front tire. As the Newton-Euler equations of motion are only valid for inertial frames, we have to derive a correlation between the acceleration a_x and a_y in the inertial frame, expressed in variables of the vehicle frame. The transformation from the vehicle frame into the inertial frame is given via the rotation matrix $R \in \mathbb{SO}(2)$ with

$$R = \begin{bmatrix} \cos(\psi) & -\sin(\psi) \\ \sin(\psi) & \cos(\psi) \end{bmatrix}. \tag{4.5}$$

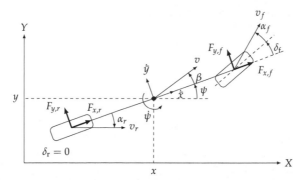

Figure 4.2. Kinetic single track model at position (x, y) with yaw angle ψ, sideslip angle β, rear and front velocity v_r, v_f, velocity at the center of mass v, rear and front steering angles $\delta_r = 0$ and δ_f, front and rear tire slip α_f and α_r, yaw rate $\dot{\psi}$, longitudinal and lateral velocity $\dot{x}$ and $\dot{y}$ in vehicle frame and the longitudinal and lateral forces with respect to the rear tire $F_{x,r}$, $F_{y,r}$ and the front tire $F_{x,f}$, $F_{y,f}$. Note that the vehicle is modeled as a rigid body affected by the respective tire forces.

We denote the acceleration and forces in inertial frame as $a = [a_x,\ a_y]^\top \in \mathbb{R}^2$ and $F = [F_x,\ F_y]^\top \in \mathbb{R}^2$. Furthermore, the acceleration and forces in the vehicle frame are given as $\bar{a} = [\bar{a}_x,\ \bar{a}_y]^\top \in \mathbb{R}^2$ and $\bar{F} = [\bar{F}_x,\ \bar{F}_y]^\top \in \mathbb{R}^2$. Note that in general, we use $(\bar{\cdot})$ to highlight that the variable is expressed in the vehicle frame. Next, we want to derive dynamical equations of the kinetic single track model. Because of the Newton-Euler equations of motion, see R. Jazar (2017)[p. 574], we know that

$$a = \frac{1}{m}F. \tag{4.6}$$

Transformed into the vehicle frame, we get $Ra = \frac{1}{m}\bar{F}$ with $\bar{F} = RF$. Furthermore, by expressing the inertial acceleration with $a = \frac{d}{dt}R^\top \begin{bmatrix} \dot{x} \\ \dot{y} \end{bmatrix}$, we get $R\frac{d}{dt}R^\top \begin{bmatrix} \dot{x} \\ \dot{y} \end{bmatrix} = \frac{1}{m}\bar{F}$, which results in

$$\begin{bmatrix} \ddot{x} - \dot{x}\,\dot{\psi} \\ \ddot{y} + \dot{y}\,\dot{\psi} \end{bmatrix} = \frac{1}{m}\bar{F}. \tag{4.7}$$

71

The overall continuous-time nonlinear dynamics of the kinetic single track model are defined as

$$
\begin{aligned}
\dot{x} &= \dot{\tilde{x}} \cos(\psi) - \dot{\tilde{y}} \sin(\psi), \\
\dot{y} &= \dot{\tilde{x}} \sin(\psi) + \dot{\tilde{y}} \cos(\psi), \\
\dot{\psi} &= \dot{\psi}, \\
\ddot{\tilde{x}} &= \dot{\psi}\, \dot{\tilde{y}} + \frac{1}{m}\, (F_{x,f} + F_{x,r}), \\
\ddot{\tilde{y}} &= -\dot{\psi}\, \dot{\tilde{x}} + \frac{1}{m}\, (F_{y,f} + F_{y,r}), \\
\ddot{\psi} &= \frac{1}{I_z}\, (l_f\, F_{y,f} - l_r\, F_{y,r}),
\end{aligned}
\tag{4.8}
$$

where m refers to the mass of the vehicle and I_z denotes the moment of inertia around the z-axis. Owing to the linear tire characteristic being only valid up to 0.4g, we use the *nonlinear Pacejka Magic Formula* to model the *lateral tire forces*, see H. Pacejka (2012). Thus, we get

$$
\begin{aligned}
F_{y,f} &= -D_f \sin(C_f \arctan(B_f\, \alpha_f - E_f\, (B_f\, \alpha_f - \arctan(B_f\, \alpha_f)))), \\
F_{y,r} &= -D_r \sin(C_r \arctan(B_r\, \alpha_r - E_r\, (B_r\, \alpha_r - \arctan(B_r\, \alpha_r)))),
\end{aligned}
\tag{4.9}
$$

with the tire slip angles defined as

$$
\alpha_f = \arctan \frac{\sin(\delta_f)\dot{\tilde{x}} + \cos(\delta_f)(\dot{\tilde{y}} + l_f\dot{\psi})}{|\cos(\delta_f)\dot{\tilde{x}} - \sin(\delta_f)(\dot{\tilde{y}} + l_f\dot{\psi})|}, \qquad \alpha_r = \arctan \frac{\dot{\tilde{y}} - l_r\dot{\psi}}{|\dot{\tilde{x}}|}.
\tag{4.10}
$$

A graphic illustration of Pacejka's Magic Formula is given in Figure 4.3. For modeling the longitudinal forces (front and rear tire $F_{x,r}$ and $F_{x,f}$), we impose the following simplifications. We neglect the tire inclination, we assume a constant load force, and we assume a direct transfer of the wheel torque to the road (no slip). Hence, the longitudinal forces at the rear and front tire are defined as

$$
\begin{aligned}
F_{x,f} &= -F_{x,f}^{f} + \frac{M}{R}, \\
F_{x,r} &= -F_{x,r}^{f} + \frac{M}{R},
\end{aligned}
\tag{4.11}
$$

where M refers to the transferred torque over the wheel radius R as a lever arm. The forces $F_{x,f}^{f}$ and $F_{x,r}^{f}$ reflect the velocity and load force dependent on friction forces defined as

$$
\begin{aligned}
F_{x,f}^{f} &= f(\dot{\tilde{x}})\, F_{l,f}, \\
F_{x,r}^{f} &= f(\dot{\tilde{x}})\, F_{r,f},
\end{aligned}
\tag{4.12}
$$

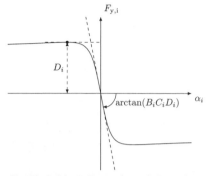

Figure 4.3. Depicted is Pacejka's Magic Formula with B_i as the stiffness factor, C_i as the shape factor and D_i as the peak value, where the variable i belongs to either the front axle $i = f$ or rear axle $i = r$. For more details, see H. Pacejka (2012)[Section 4.3.1].

where the friction force is $f(\dot{x}) = \text{sign}(\dot{x}) \ f_0 + f_1 \ \dot{x} + \text{sign}(\dot{x}) \ f_4 \ \dot{x}^4$ with $f_0, f_1, f_4 \in \mathbb{R}$. Moreover, the load forces for the front and rear axle are defined as $F_{l,f} = \frac{m \, l_r \, g}{l_f + l_r}$ and $F_{l,r} = \frac{m \, l_f \, g}{l_f + l_r}$.

4.2 Adapted relaxed barrier function MPC algorithms

In the following, we modify the rbMPC trajectory tracking and path following open-loop optimal control problem proposed in Chapter 2 for the application to nonlinear systems. Basically, two modifications are needed. First, we follow the ideas of A. Sabelhaus et al. (2018), modifying the weighting of the input from $||u(k) - u_{\text{ref}}(k)||_R^2$ into the well-known Δu-formulation with $||\Delta u||_R^2$. Second, we follow the ideas of P. Falcone et al. (2008), linearizing the nonlinear system around a given reference trajectory $\{\xi_{\text{ref}}(k), u_{\text{ref}}(k)\}_{k=0...\infty}$ and using the linearized linear time-varying (LTV) system as a prediction model for the rbMPC algorithms. More precisely, we consider a nonlinear model $f : \mathbb{R}^n \times \mathbb{R}^m \to \mathbb{R}^n$ that is defined as

$$\xi(k+1) = f(\xi(k), u(k)), \qquad \xi(0) = \xi_0, \qquad (4.13)$$

where $\xi(k) \in \mathbb{R}^n$ and $u(k) \in \mathbb{R}^m$ refer to the system state and control input at sampling instant $k \in \mathbb{N}$, respectively. We approximate the nonlinear system (4.13) by successive linearization around the reference trajectory. This results in the following LTV system

$$\Delta \xi(k+1) = A(k)\Delta \xi(k) + B(k)\Delta u(k), \tag{4.14}$$

where the time-varying matrices $A(k)$ and $B(k)$ are defined as

$$A(k) = \frac{\partial f}{\partial \xi}\Big|_{(\xi_{\text{ref}}(k), u_{\text{ref}}(k))}, \qquad B(k) = \frac{\partial f}{\partial u}\Big|_{(\xi_{\text{ref}}(k), u_{\text{ref}}(k))} \tag{4.15}$$

and $\Delta \xi(k) = \xi(k) - \xi_{\text{ref}}(k)$, $\Delta u(k) = u(k) - u_{\text{ref}}(k)$. Note that this approximation is only valid if we stay close to the reference trajectory, which we can assume without the loss of generality. The dynamic behavior of the LTV system (4.14) i-th step ahead is defined as

$$\Delta \xi(k+i+1) = A_i(k)\Delta \xi(k+i) + B_i(k)\Delta u(k+i) \tag{4.16}$$

with $i \in \mathbb{N}$ and the time-varying matrices $A_i(k)$ and $B_i(k)$ defined as

$$A_i(k) = \frac{\partial f}{\partial \xi}\Big|_{(\xi_{\text{ref}}(k+i), u_{\text{ref}}(k+i))}, \qquad B_i(k) = \frac{\partial f}{\partial u}\Big|_{(\xi_{\text{ref}}(k+i), u_{\text{ref}}(k+i))}. \tag{4.17}$$

To obtain the original system state i-th step ahead, we rewrite eq. (4.16) as

$$\xi(k+1+i) = A_i(k)\xi(k+i) + B_i(k)u(k+i) + e_i(k) \tag{4.18}$$

with $e_i(k) = \xi_{\text{ref}}(k+1+i) - A_i(k)\xi_{\text{ref}}(k+i) - B_i(k)u_{\text{ref}}(k+i)$. We like to end this section with some remark about the calculation of the reference input.

Remark 24. In general, the calculation of the reference control input $u_{\text{ref}}(k+i)$, $i = 0, \ldots, N-1$ along the prediction horizon $N \in \mathbb{N}_+$ is a non-trivial task. We like to mention that in our cases the reference input is given a priori which is either calculated based on the nonlinear kinematic single track model for the maneuvers in Subsection 4.4.1 and 4.4.2 or given by measurement data for the maneuver in Subsection 4.4.3.

4.2.1 Trajectory tracking with relaxed barrier function MPC

Based on the LTV system (4.18) and the Δu formulation, see A. Sabelhaus et al. (2018), we adapt the non-relaxed open-loop optimal control problem to

$$
\begin{aligned}
\min_{\Delta u} \quad & \sum_{i=0}^{N-1} ||\xi_i - \xi_{\text{ref}}(k+i)||_Q^2 + ||\Delta u_i||_R^2 + ||\xi_N - \xi_{\text{ref}}(k+N)||_P^2 \\
\text{s.t.} \quad & \xi_{i+1} = A_i(k)\xi_i + B_i(k)u_i + e_i(k), \quad \xi_0 = \xi(k), \\
& \xi_{\min} \leq \xi_i \leq \xi_{\max}, \ i = 0, \ldots, N-1, \\
& u_{\min} \leq u_i \leq u_{\max}, \ i = 0, \ldots, N-1, \\
& \Delta u_{\min} \leq \Delta u_i \leq \Delta u_{\max}, \ i = 0, \ldots, N-1, \\
& u_i = u_{i-1} + \Delta u_i, \ i = 0, \ldots, N-1, \\
& u_{-1} = u(k-1),
\end{aligned}
\tag{4.19}
$$

where the open-loop optimal control problem is parametrized with the reference trajectory $\{\xi_{\text{ref}}(k+i)\}_{i=0,\ldots,N}$, $\{u_{\text{ref}}(k+i)\}_{i=0,\ldots,N-1}$, the previous control input $u(k-1) \in \mathbb{R}^m$, and the current system state $\xi(k) \in \mathbb{R}^n$. Moreover, $Q \in \mathbb{S}_{++}^n$, $R \in \mathbb{S}_{++}^m$ are user-defined tuning matrices and $P \in \mathbb{S}_{++}^n$. The decision variables are defined as $\Delta u = \{\Delta u_0^* \ \ldots \ \Delta u_{N-1}^*\}$. In the following, we transform the open-loop optimal control problem (4.19) into an open-loop optimal control problem based upon relaxed barrier functions, according to Chapter 2. Moreover, we formulate the open-loop optimal control problem (4.19) in vector-matrix form, which results in a parametric convex optimization program. In general, there exist two vector-matrix formulations of the open-loop optimal control problem (4.19), the *sparse* and the *condensed* formulation. Based on the evaluations of T. Schanz (2018), we know that the most expensive task is the calculation of the Hessian and not the optimization itself. Thus, we decide to choose the condensed formulation with optimization vector $\Delta U = [\Delta u_0^{*\top} \ \ldots$ $\Delta u_{N-1}^{*\top}]^\top \in \mathbb{R}^{Nm}$ and parameter vector $\zeta = [\xi_0^\top \ u_{-1}^\top \ u_{\text{ref}}(k)^\top \ \xi_{\text{ref}}(k)^\top \ \ldots$ $u_{\text{ref}}(k+N-1)^\top \xi_{\text{ref}}(k+N)^\top]^\top \in \mathbb{R}^{(N+1)(n+m)+n}$. In the following, we make use of a condensed formulation such that the equality constraints are transferred into the cost. Moreover, the inequality constraints are written into the cost by the usage of relaxed barrier functions. Hence, we result in the following unconstrained parametric convex optimization program

$$
\min_{\Delta U} \Delta U^\top H \Delta U + F \Delta U + s + \epsilon \hat{B}_G(\Delta U, k),
\tag{4.20}
$$

where $H \in \mathbb{S}_{++}^{Nm}$, $F \in \mathbb{R}^{1 \times Nm}$, $s \in \mathbb{R}$ and $\hat{B}_{\mathrm{G}} : \mathbb{R}^{Nm} \times \mathbb{N} \to \mathbb{R}_+$ being a gradient recentered relaxed barrier function

$$\hat{B}_{\mathrm{G}}(\Delta U, k) = \sum_{i=1}^{q} \hat{B}(-G^i \Delta U + r + d^i(k)) + \ln(d^i(k)) - \frac{G^i \Delta U}{d^i(k)} \qquad (4.21)$$

with $G \in \mathbb{R}^{q \times Nm}$ and $d(k) \in \mathbb{R}_{++}^{q}$ with $q = N(q_{\mathrm{x}} + q_{\mathrm{u}} + q_{\Delta \mathrm{u}})$ where $q_{\mathrm{x}}, q_{\mathrm{u}}$ and $q_{\Delta \mathrm{u}} \in \mathbb{N}$ refer to the number of state, input, and Δ input constraints. Moreover, the relaxed barrier function $\hat{B}(\cdot)$ is defined according to Chapter 2 Definition 1 and G^i refers to the i-th row of a general matrix G. For the actual realization of the cost and the inequality matrices, we refer to Appendix C.1. The control input $u(k)$ at sampling instant $k \in \mathbb{N}$ is obtained from the optimal solution $\Delta U^*(k) \in \mathbb{R}^{Nm}$ of the parametric convex optimization program (4.20). More precisely, we have that

$$u(k) = u(k-1) + \Pi_0 \Delta U^*(k), \qquad (4.22)$$

where $\Pi_0 = \begin{bmatrix} I_m & 0_m & \dots & 0_m \end{bmatrix} \in \mathbb{R}^{m \times Nm}$ refers to a selection matrix. Finally, the rbMPC trajectory tracking algorithm consists of the following steps

---

Step 1: Measure current system state $\xi(k)$, previous control input $u(k-1)$ and get the reference trajectory $\{\xi_{\mathrm{ref}}(k+i)\}_{i=0,\dots,N}$, $u_{\mathrm{ref}}(k+i)\}_{i=0,\dots,N-1}$ along the prediction horizon $N \in \mathbb{N}_+$.

Step 2: Solve the unconstrained parametric convex optimization program (4.20) by Newton's method, see Subsection 4.2.3.

Step 3: Apply the optimal solution $\Delta U^*(k)$ with $u(k) = u(k-1) + \Pi_0 \Delta U^*(k)$ to the system (4.13).

Step 4: Go back to Step 1 and repeat for the next sampling instant $k+1$.

Figure 4.4. Steps of the rbMPC trajectory tracking algorithm.

These four steps are similar to the steps of a standard MPC scheme. Note that we will investigate Step 2 in Subsection 4.2.3 in detail.

4.2.2 Path following with relaxed barrier function MPC

Note that in general the quadratic program of the proposed path following rbMPC algorithm is due to the nonlinearity of the path $P(\theta)$ nonlinear. To achieve a computational more efficient algorithm, we make in the latter the following assumption.

Assumption 25. The reference path is defined by

$$r(\theta) = (\xi_{\text{ref}}(\theta), u_{\text{ref}}(\theta)) \tag{4.23}$$

with $\xi_{\text{ref}}(\theta) = \xi_{\text{ref}}(k) + (\theta - k)(\xi_{\text{ref}}(k+1) - \xi_{\text{ref}}(k))$ for $k \leq \theta < k+1$ and $u_{\text{ref}}(\theta) = u_{\text{ref}}(k) + (\theta - k)(u_{\text{ref}}(k+1) - u_{\text{ref}}(k))$ for $k \leq \theta < k+1$ for all $(\theta, k) \in \mathbb{R} \times \mathbb{N}$.

Assumption 25 represents a linear approximation of the reference state and input at two consecutive sampling instants. Thus, the reference path (4.23) is piece wise defined depending on the sampling instant $k \in \mathbb{N}$. The system state and control input open-loop predictions are obtained by

$$\begin{aligned} \xi_{\text{ref}}(\theta_j) &= \xi_{\text{ref}}(k_j) + (\xi_{\text{ref}}(k_j + 1) - \xi_{\text{ref}}(k_j))(\theta_j - k_j), \; j = 0, \ldots, N, \\ u_{\text{ref}}(\theta_j) &= u_{\text{ref}}(k_j) + (u_{\text{ref}}(k_j + 1) - u_{\text{ref}}(k_j))(\theta_j - k_j), \; j = 0, \ldots, N-1 \end{aligned} \tag{4.24}$$

with θ_j being the open-loop predictions of the path parameter and $k_j = \lfloor (\theta_j^*(k-1)) \rfloor$ where $\theta^*(k-1) = \{\theta_1^*(k-1), \ldots, \theta_N^*(k-1), \theta_N^*(k-1) + 1\}$ defines the shifted open-loop predictions of the path parameter at the previous sampling instant. Furthermore, $\lfloor x \rfloor = \max\{i \in \mathbb{Z} | i \leq x\}$ for $x \in \mathbb{R}$. Based on the well-known Δu formulation, see A. Sabelhaus et al. (2018), and an additional regularization term of the path parameters virtual input, we adapt the non-relaxed open-loop optimal control problem for the path following problem to nonlinear systems by successive linearization, see Section 4.2. More precisely, we have

$$\begin{aligned} \min_{\Delta u, v} \quad & \sum_{i=0}^{N-1} ||\xi_i - \xi_{\text{ref}}(\theta_i)||_Q^2 + ||\Delta u_i||_R^2 + ||v_i||_T^2 + ||\xi_N - \xi_{\text{ref}}(\theta_N)||_P^2 \\ \text{s.t.} \quad & \xi_{i+1} = A(\theta_i^*(k-1))\xi_i + B(\theta_i^*(k-1))u_i + e_i(\theta_i^*(k-1)), \quad \xi_0 = \xi(k), \\ & \theta_{i+1} = \theta_i + 1 + v_i, \quad \theta_0 = \theta(k), \\ & \xi_{\text{ref}}(\theta_i) = \xi_{\text{ref}}(k_i) + (\xi_{\text{ref}}(k_i + 1) - \xi_{\text{ref}}(k_i))(\theta_i - k_i), \quad k_i = \lfloor \theta_i^*(k-1) \rfloor, \\ & v_{\min} \leq v_i \leq v_{\max}, \; i = 0, \ldots, N-1, \\ & \xi_{\min} \leq \xi_i \leq \xi_{\max}, \; i = 0, \ldots, N-1, \\ & u_{\min} \leq u_i \leq u_{\max}, \; i = 0, \ldots, N-1, \\ & \Delta u_{\min} \leq \Delta u_i \leq \Delta u_{\max}, \; i = 0, \ldots, N-1, \\ & u_i = u_{i-1} + \Delta u_i, \; i = 0, \ldots, N-1, \\ & u_{-1} = u(k-1), \end{aligned}$$

$$\tag{4.25}$$

where the open-loop optimal control problem is parametrized with the reference trajectory $\{\xi_{\text{ref}}(k_i)\}_{i=0,\ldots,N}$, $\{u_{\text{ref}}(k_i)\}_{i=0,\ldots,N-1}$, the previous control input

$u(k-1) \in \mathbb{R}^m$, the current system state $\zeta(k) \in \mathbb{R}^m$, the current path parameter $\theta(k) \in \mathbb{R}$, and the shifted open-loop predictions of the path parameter at the previous sampling instant $\theta^*(k-1) = \{\theta_1^*(k-1), \ldots, \theta_N^*(k-1), \theta_N^*(k-1)+1\}$. Furthermore, $\lfloor x \rfloor = \max\{i \in \mathbb{Z} | i \leq x\}$ with $x \in \mathbb{R}$ and $Q \in \mathbb{S}_{++}^n$, $R \in \mathbb{S}_{++}^m$, $T \in \mathbb{S}_{++}$ being user-defined tuning matrices and the terminal cost matrix $P \in \mathbb{S}_{++}^n$. The decision variables are defined as $\boldsymbol{\Delta u} = \{\Delta u_0^* \ldots \Delta u_{N-1}^*\}$ and $\boldsymbol{v} = \{v_0^* \ldots v_{N-1}^*\}$. Note that the path-varying matrices $A(\theta_i^*(k-1))$ and $B(\theta_i^*(k-1))$ are obtained by linearization of the nonlinear dynamics (4.1) around $(\zeta_{\mathrm{ref}}(\theta_i^*(k-1)), u(\theta_i^*(k-1)))$ i-th step ahead with $i \in \mathbb{N}$. More precisely, we have

$$
\begin{aligned}
A(\theta_i^*(k-1)) &= \frac{\partial f}{\partial x}(\zeta_{\mathrm{ref}}(\theta_i^*(k-1)), u_{\mathrm{ref}}(\theta_i^*(k-1))), \\
B(\theta_i^*(k-1)) &= \frac{\partial f}{\partial u}(\zeta_{\mathrm{ref}}(\theta_i^*(k-1)), u_{\mathrm{ref}}(\theta_i^*(k-1))).
\end{aligned}
\tag{4.26}
$$

Next, we follow along the lines of Subsection 4.2.1 and reformulate the open-loop optimal control problem into a *condensed* vector-matrix form with the optimization vector being denoted as $w = [\Delta u_0^\top \ldots \Delta u_{N-1}^\top \ v_0 \ldots v_{N-1}]^\top \in \mathbb{R}^{N(m+1)}$ and the parameter vector being defined as $\zeta = [\zeta_0^\top \ u_{-1}^\top \ u_{\mathrm{ref}}(k_0)^\top \ \theta_0^*$ $\zeta_{\mathrm{ref}}(k_0)^\top \ \ldots \ u_{\mathrm{ref}}(k_{N-1})^\top \ \theta_N^* \ \zeta_{\mathrm{ref}}(k_N)^\top]^\top \in \mathbb{R}^{(N+1)(n+m)+N}$. By writing the equality constraints and especially the inequality constraints with relaxed barrier functions into the cost, we have the following unconstrained parametric convex optimization program

$$
\min_w w^\top H w + F w + s + \epsilon \hat{B}_{\mathrm{G}}(w, \theta) + \epsilon \hat{B}_{\mathrm{G},v}(w)
\tag{4.27}
$$

with the cost matrices $H \in \mathbb{S}_{++}^{N(m+1)}$, $F \in \mathbb{R}^{1 \times N(m+1)}$, $s \in \mathbb{R}$, $\hat{B}_{\mathrm{G},v} : \mathbb{R}^{N(m+1)} \to \mathbb{R}_+$ and $\hat{B}_{\mathrm{G}} : \mathbb{R}^{N(m+1)} \times \mathbb{R} \to \mathbb{R}_+$ being the gradient recentered and relaxed barrier function, see Chapter 2, which is defined as

$$
\begin{aligned}
\hat{B}_{\mathrm{G}}(w, \theta) &= \sum_{i=1}^{q_{\mathrm{x}}+q_{\mathrm{u}}+q_{\Delta u}} \hat{B}(-G^i w + r + d^i(\theta)) + \ln(d^i(\theta)) - \frac{G^i w}{d^i(\theta)}, \\
\hat{B}_{\mathrm{G},v}(w) &= \sum_{i=1}^{q_{\mathrm{v}}} \hat{B}(-G_{\mathrm{v}}^i w + d_{\mathrm{v}}^i) + \ln(d_{\mathrm{v}}^i) - \frac{G_{\mathrm{v}}^i w}{d_{\mathrm{v}}^i},
\end{aligned}
\tag{4.28}
$$

with the inequality matrices $G \in \mathbb{R}^{q \times N(m+1)}$, $d_{\mathrm{v}} \in \mathbb{R}^{q_{\mathrm{v}}}$, $d(\theta) \in \mathbb{R}_{++}^q$, and $q = q_{\mathrm{x}} + q_{\mathrm{u}} + q_{\Delta u} + q_{\mathrm{v}}$ with $q_{\mathrm{x}}, q_{\mathrm{u}}, q_{\Delta u}, q_{\mathrm{v}} \in \mathbb{N}$ denoting the number of state, input, Δ input, and virtual input constraints, respectively. Moreover G^i refers to the i-th row of a general matrix G. For the actual realization of the cost and inequality matrices, we refer to Appendix C.2. The control input $u(k)$ and the virtual control input $v(k)$ at sampling instant $k \in \mathbb{N}$ are obtained from the optimal solution $w^*(k) \in \mathbb{R}^{N(m+1)}$ of the parametric convex optimization program (4.27). More precisely, we have

$$
\begin{aligned}
u(k) &= u(k-1) + \Pi_0 w^*(k), \\
v(k) &= \Pi_1 w^*(k),
\end{aligned}
\tag{4.29}
$$

where $\Pi_0 = \begin{bmatrix} I_m & 0_m & \dots & 0_m \end{bmatrix} \in \mathbb{R}^{m \times N(m+1)}$ and $\Pi_1 = \begin{bmatrix} 0 & 0 & \dots & 1 & 0 & \dots & 0 \end{bmatrix} \in \mathbb{R}^{N(m+1)}$ are selection matrices. Finally, the overall rbMPC path following algorithm consists of the following steps

---

Step 1: Measure the current system state $\xi(k)$, the previous control input $u(k-1)$, get the current path parameter $\theta(k)$ and the reference trajectory $\{\xi_{\text{ref}}(\theta_i^*(k-1))\}_{i=0,\dots,N}$, $\{u_{\text{ref}}(\theta_i^*(k-1))\}_{i=0,\dots,N-1}$ at the shifted open-loop predictions of the path parameter at the previous sampling instant $\theta^*(k-1) = \{\theta_1^*(k-1), \dots, \theta_N^*(k-1), \theta_N^*(k-1) + 1\}$.

Step 2: Solve the unconstrained parametric convex optimization program (4.27) by Newton's method, see Subsection 4.2.3.

Step 3: Apply the optimal solution $w^*(k)$ with $u(k) = u(k-1) + \Pi_0 w^*(k)$ to the system (4.13) and apply the virtual control input $v(k) = \Pi_1 w^*(k)$ to the system $\theta(k+1) = \theta(k) + v(k)$.

Step 4: Go back to Step 1 and repeat for sampling instant $k+1$.

Figure 4.5. Steps of the rbMPC path following algorithm.

4.2.3 Newton method with warm start solution

In this subsection, we discuss Step 2 of the algorithms (4.4) and (4.5) dealing with the numerical solution of the unconstrained convex parametric optimization programs (4.20) and (4.27). Consider a general optimizer $r(k) \in \mathbb{R}^r$ of the unconstrained convex parametric optimization program, e.g. in case of the rbMPC tracking algorithm $r(k) = \Delta U(k) \in \mathbb{R}^{Nm}$ with $r = Nm$ and in case of the rbMPC path following algorithm $r(k) = w(k) \in \mathbb{R}^{N(m+1)}$ with $r = N(m+1)$. Moreover, we define a general initial condition $\zeta(k) \in \mathbb{R}^s$, e.g. in case of the rbMPC trajectory tracking algorithm we have $\zeta(k) \in \mathbb{R}^{(N+1)(n+m)+n}$ with $s = (N+1)(n+m) + n$ and in case of the rbMPC path following algorithm, we get $\zeta(k) \in \mathbb{R}^{(N+1)(n+m)+N}$ with $s = (N+1)(n+m) + N$. Furthermore, $\Pi \in \mathbb{R}^{\tilde{n}} \times \mathbb{R}^r$ is a general selection matrix which chooses the current system control input $u(k)$ out of the general optimizer $r(k)$. Using the notation introduced above, we can summarize the closed-loop dynamics to

$$
\begin{aligned}
\xi(k+1) &= f(\xi(k), \Pi r(k)), \quad \xi(0) = \xi_0, \\
r(k+1) &= \Phi^{i_T(k)}(r(k), \zeta(k)), \quad r(0) = r_0,
\end{aligned}
\tag{4.30}
$$

where $\Phi^{i_T(k)} : \mathbb{R}^r \times \mathbb{R}^s \to \mathbb{R}^r$ denotes the optimization procedure repeated $i_T(k) \in \mathbb{N}_+$ times, defined recursively as

$$
\Phi^0(r(k), \zeta(k)) = \Gamma_w(r(k), \zeta(k)),
\tag{4.31}
$$

$$
\Phi^i(r(k), \zeta(k)) = \Gamma_o(\Phi^{i-1}(r(k), \zeta(k)), \zeta(k+1)),
\tag{4.32}
$$

with $\Gamma_w : \mathbb{R}^r \times \mathbb{R}^s \to \mathbb{R}^r$ corresponds to a suitable warm-start solution and the operator $\Gamma_o : \mathbb{R}^r \times \mathbb{R}^s \to \mathbb{R}^r$ refers to a suitable optimization method, for example Newton's method. A warm-start solution is chosen as the shift operator for the rbMPC trajectory tracking algorithm as

$$
\Gamma_w(r(k), \zeta(k)) = [\Delta u_1^*(k-1)^\top \ \dots \ \Delta u_{N-1}^*(k-1)^\top \ \Delta u_{N-1}^*(k-1)^\top]^\top
\tag{4.33}
$$

and for the rbMPC path following algorithm as

$$
\begin{aligned}
\Gamma_w(r(k), \zeta(k)) = [&\Delta u_1^*(k-1)^\top \ \dots \ \Delta u_{N-1}^*(k-1)^\top \\
&\Delta u_{N-1}^*(k-1)^\top \ 0 \ \dots \ 0]^\top.
\end{aligned}
\tag{4.34}
$$

Because of the open-loop optimal control problems (4.20) and (4.27) being *unconstrained* optimization problems that are twice continuously differentiable and strictly convex, we can apply standard numerical optimization techniques such as the Newton method. The following paragraph deals with Newton's method to solve a convex and unconstrained optimization problem and summarizes the most important findings of S. Boyd and L. Vandenberghe (2004)[Section 9]. Consider an unconstrained optimization problem

$$J_N^*(\zeta) = \min_r J_N(r, \zeta),\qquad (4.35)$$

where the value function $J_N : \mathbb{R}^r \times \mathbb{R}^s \to \mathbb{R}$ is strongly convex and twice continuously differentiable. By strongly convex, we mean that there exists an $m > 0$ such that $\nabla_r^2 J_N(r, \zeta) \geq mI$ which is the case for the value function of the open-loop optimal control problem (4.20) and (4.27). Since $J_N(r, \zeta)$ is differentiable and strictly convex, a necessary and sufficient condition for r^* to be the optimal solution is $\nabla_r J_N(r^*, \zeta) = 0$. In general, the optimal solution r^* cannot be determined analytically. Thus, we aim to solve the unconstrained optimization problem (4.35) in terms of an iterative algorithm that computes an infinite sequence $r = \{r^0, r^1, \ldots, r^\infty\}$ with $\lim_{i \to \infty} J_N(r^i, \zeta) = J_N(r^*, \zeta)$. More precisely, we define the iterative algorithm to be of the form

$$r^{i+1} = r^i + \alpha^i p^i, \qquad r^0 \in \mathbb{R}^r, \qquad (4.36)$$

where $\alpha^i \in \mathbb{R}_+$ and $p^i \in \mathbb{R}^n$ denote the step length and the search direction at the iteration number $i \in \mathbb{N}_+$. We call the search direction a descent direction if

$$\nabla_r J_N(r^i, \zeta)^\top p^i < 0. \qquad (4.37)$$

The definition of the descent direction originates from the first-order Taylor approximation of $J_N(r^i + v, \zeta)$ around r^i that leads to $\bar{J}_N(r^i + v, \zeta) = J_N(r^i, \zeta) + \nabla_r J_N(r^i, \zeta)^\top v$. The part $\nabla_r J_N(r^i, \zeta)^\top v$ is the directional derivative. Obviously, the step v is a descent direction if $\nabla_r J(r^i, \zeta)^\top v < 0$. The search direction within Newton's method is defined as

$$p_N^i = -\nabla_r^2 J_N(r^i, \zeta)^{-1} \nabla_r J_N(r^i, \zeta), \qquad (4.38)$$

which is, due to the strict convexity, a descent direction. More precisely, it holds that $-\nabla_r J_N(r^i, \zeta)^\top \nabla_r^2 J_N(r^i, \zeta)^{-1} \nabla_r J_N(r^i, \zeta) < 0$. The search direction by Newton's method, also called the Newton step, originates from the idea of computing the optimal search direction for a second-order Taylor approximation of $J_N(r + d, \zeta)$ around r, which leads to

$$\hat{J}_N(r^i + d, \zeta) = J_N(r^i, \zeta) + \nabla_r J_N(r^i, \zeta)^\top d + \frac{1}{2} d^\top \nabla_r^2 J_N(r^i, \zeta) d \qquad (4.39)$$

By applying the first-order optimality condition to (4.39), we obtain the Newton step. More precisely, we have

$$p_N^i = \arg\min_d \hat{J}_N(r^i + d, \zeta) = -\nabla_r^2 J_N(r^i, \zeta)^{-1} \nabla_r J_N(r^i, \zeta). \qquad (4.40)$$

By inserting the Newton step p_N^i into eq. (4.39) and considering

$$
\begin{aligned}
J_N(r^i, \zeta) - \hat{J}_N(r^i + p_N^i, \zeta) &= \frac{1}{2} \nabla_r J_N(r^i, \zeta)^\top \nabla_r^2 J_N(r^i, \zeta)^{-1} \nabla_r J_N(r^i, \zeta) \\
&= \frac{1}{2} \lambda(r^i, \zeta)^2,
\end{aligned} \qquad (4.41)
$$

we get an estimate for the suboptimality $J_N(r, \zeta) - J_N^*(r, \zeta)$. The quantity $\lambda(r^i, \zeta)$ is called Newton decrement at r^i and ζ. It plays an important role in the analysis of Newton's method, see S. Boyd and L. Vandenberghe (2004)[Section 9.5], which will be used as a stopping criteria for Newton's method, see Algorithm 4.7.

The step size in Newton's method is chosen according to the backtracking line search given below.

Descent direction p^i for J_N, $\alpha \in (0, 0.5)$ and $\beta \in (0, 1)$
function BACKTRACKING LINE SEARCH(p^i, α, β)
$\alpha^0 = 1$;
    while $J_N(r^i + \alpha^i p^i, \zeta) > J_N(r^i, \zeta) + \alpha \alpha^i \nabla J_N(r^i, \zeta)^\top p^i$ **do**
$\alpha^{i+1} = \beta \alpha^i$
    end while
end function

Figure 4.6. Backtracking line search of Newton's method, see S. Boyd and L. Vandenberghe (2004).

$r^0 \in \mathbb{R}^r$
function NEWTON'S METHOD(r^0)
    $\epsilon_N > 0, 0 < \alpha < 0.5, 0 < \beta < 1, i = 0$
    while true **do**
        Calculate Newton step $p_N^i = -\nabla_r^2 J_N(r^i, \zeta)^{-1} \nabla_r J_N(r^i, \zeta)$
        Calculate Newton decrement $\lambda(r^i, \zeta)^2 = -\nabla_r J_N(r^i, \zeta)^\top p_N^i$
        if $\lambda(r^i, \zeta)^2 \leq \epsilon_N$ **then**
            $r^* = r^i$
            return r^*
        end if
        function BACKTRACKING LINE SEARCH(p_N^i, α, β)
        end function
        $r^{i+1} = r^i + \alpha^i p_N^i$
        $i = i + 1$
    end while
end function

Figure 4.7. Summary of Newton's method, see S. Boyd and L. Vandenberghe (2004).

The backtracking line search starts with unit step size $\alpha^i = 1$ and then sequentially reduces the step size by factor $\beta \in (0, 1)$ until the inequality

$$J_N(r^i + \alpha^i p^i, \zeta) > J_N(r^i, \zeta) + \alpha \alpha^i \nabla_r J_N(r^i, \zeta)^\top p^i \tag{4.42}$$

is satisfied for some $\alpha \in (0, 0.5)$. This inequality can always be satisfied for a small enough step size α^i given that the Newton step p_N^i is a descent direction and thus we have, based on a first-order Taylor approximation, that $J_N(r^i, \zeta) + \alpha^i \nabla_r J_N(r^i, \zeta)^\top p^i < J_N(r^i, \zeta) + \alpha t \nabla_r J_N(r^i, \zeta)^\top p^i$. The performance of the backtracking line search depends mainly on the choice of the parameters α and β. The parameter α describes the decrease within the function $J_N(r, \zeta)$ compared with its linear approximation. The parameter β influences how crude the line search is. According to S. Boyd and L. Vandenberghe (2004), the parameter α is typically chosen between 0.01 and 0.3. The parameter β is chosen to be between 0.1 and 0.8. The overall Newton's method is summarized in Figure 4.7, see S. Boyd and L. Vandenberghe (2004).

4.3 Benchmark controller

In Subsections 4.2.1 and 4.2.2, we derived rbMPC algorithms for trajectory tracking and path following for nonlinear systems. Because of the internal single track model used by the respective rbMPC schemes, the lateral and longitudinal dynamics are implicitly coupled. Thus, the developed rbMPC schemes belong to the class of coupled lateral and longitudinal control. Next, we will derive a benchmark controller that uses a decoupled lateral and longitudinal controller mainly used in industry, see C. Olsson (2015); R. Attia, R. Orjuela and M. Basset (2014)). Note that that the decoupling of the longitudinal and lateral dynamics constitutes its main simplification. Following along the ideas of R. Coulter (1992), J. Morales et al. (2009) and M. Samuel, M. Hussein and M. Binti (2016), we propose a pure pursuit algorithm for *lateral control*. The pure pursuit controller, as the word "pursuit" implies, tries to follow/chase a goal point that lies in front of the vehicle in the driving direction. A geometric explanation of the pure pursuit algorithm is given below.

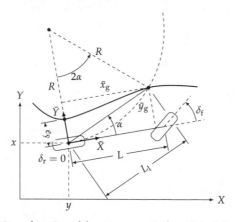

Figure 4.8. Geometric explanation of the pure pursuit algorithm at the example of a kinematic single track model at position (x, y), goal position $(\bar{x}, \bar{y})$, angle α, radius R, length of the vehicle L, look-ahead distance L_l, front and rear steering angles $\delta_r = 0$, and lateral error $e_{\bar{y}}$.

Here, (x, y) denotes the coordinates of the center of the rear tire in the inertial frame (X, Y), which refers to the vehicle's reference point. The goal point $(\bar{x}_g, \bar{y}_g)$ is depicted in the vehicle frame $(\bar{X}, \bar{Y})$ which is connected via a circle of radius R with the vehicle's reference point. The distance between the goal and the vehicle's reference point is the so-called look ahead distance L_l, where the angle α is the difference angle between the heading of the single-track model and the connection line between the goal and the vehicle's reference point. The front and rear steering angle are depicted as δ_r, δ_f, where we assume that $\delta_r = 0$. The overall length of the single-track model refers to L, and the lateral error is defined as $e_{\tilde{y}}$. From Figure 4.8, we can conclude the basic functionality of the pure pursuit algorithm. Based on Ackermann steering geometry, the pure pursuit algorithm calculates the desired curvature

$$\kappa = \frac{2\sin(\alpha)}{L_l} \tag{4.43}$$

to get to the goal point $(\bar{x}_g, \bar{y}_g)$. Finally, the desired curvature is transformed into a front steering angle with

$$\delta = \tan^{-1}\left(\frac{2L\sin(\alpha)}{L_l}\right). \tag{4.44}$$

More precisely, the pure pursuit algorithm consists of the following steps

Step 1: Determine the current vehicle's reference point (x, y) in the inertial frame (X, Y).

Step 2: Find the path point closest to the vehicle's reference point.

Step 3: Find the goal point depending on the look ahead distance L_l.

Step 4: Transform the goal point into vehicle coordinates $(\bar{x}_g, \bar{y}_g)$.

Step 5: Calculcate the curvature and apply the front steering angle δ_f to the vehicle according to eq.. (4.43) and (4.44)

Step 6: Go to Step 1.

Figure 4.9. Steps of the pure pursuit algorithm.

To be able to control the vehicle's lateral and longitudinal dynamics, we have to design a longitudinal controller. Hence, we will propose a *longitudinal controller*. Following the ideas of G. Pannocchia and J. Rawlings (2001), we propose an output set point tracking algorithm based on an linear quadratic regulator (LQR) controller in velocity form. Consider a LTI discrete-time system

$$\begin{aligned} \xi(k+1) &= A\xi(k) + Bu(k), \qquad \xi(0) = \xi_0 \\ y(k) &= C\xi(k), \end{aligned} \tag{4.45}$$

where $\xi(k) \in \mathbb{R}^n$, $u(k) \in \mathbb{R}^m$, and $y(k) \in \mathbb{R}^p$ refers to the system state, the control input, and the system output at sampling instant $k \in \mathbb{N}$, respectively.

Assumption 26. We assume that the system (4.45) is stabilizable.

Assumption 27. We assume that

$$\text{rank} \begin{bmatrix} I - A & B \\ C & 0 \end{bmatrix} = n + p. \tag{4.46}$$

Assumption 27 is needed to obtain offset-free control, see G. Pannocchia and J. Rawlings (2001). By introducing new variables

$$\Delta\xi(k) = \xi(k) - \xi(k-1), \; z(k) = y(k) - y_{\text{ref}}(k),$$
$$\Delta u(k) = u(k) - u(k-1), \tag{4.47}$$

where $y_{\text{ref}}(k)$ is assumed to be known, for the time-invariant case, we refer to G. Pannocchia and J. Rawlings (2001). Following along the time-invariant case, we can define the following augmented system

$$\tilde{\xi}(k+1) = \tilde{A}\tilde{\xi}(k) + \tilde{B}\tilde{u}(k), \qquad \tilde{\xi}(0) = \tilde{\xi}_0$$
$$z(k) = \tilde{C}\tilde{\xi}(k) \tag{4.48}$$

with $\tilde{\xi}(k) = \begin{bmatrix} \Delta\xi(k)^\top & z_k^\top \end{bmatrix}^\top$, $\tilde{u}(k) = \Delta u_k$, and the matrices

$$\tilde{A} = \begin{bmatrix} A & 0 \\ CA & I \end{bmatrix}, \quad \tilde{B} = \begin{bmatrix} B \\ CB \end{bmatrix}, \quad \tilde{C} = \begin{bmatrix} 0 & I_p \end{bmatrix}. \tag{4.49}$$

From G. Pannocchia and J. Rawlings (2001), we know that if (A, B) is stabilizable, then also the augmented system $(\tilde{A}, \tilde{B})$ is stabilizable. The regulator problem can be formulated as an infinite horizon discrete-time open-loop optimal control problem

$$\min \; \frac{1}{2} \sum_{k=0}^{\infty} \tilde{\xi}_k^\top \tilde{Q}\tilde{\xi}_k + \tilde{u}_k^\top \tilde{R}\tilde{u}_k \tag{4.50}$$
$$\text{s.t. } \tilde{\xi}_{k+1} = \tilde{A}\tilde{\xi}_k + \tilde{B}\tilde{u}_k, \quad \tilde{\xi}_0 = \tilde{\xi}(0)$$

with $\tilde{R} > 0$ and

$$\tilde{Q} = \begin{bmatrix} 0 & 0 \\ 0 & Q \end{bmatrix} \tag{4.51}$$

with $Q > 0$. The optimal solution of (4.50) is

$$\tilde{u}_k = -(\tilde{R} + \tilde{B}^\top S \tilde{B})^{-1} \tilde{B}^\top S \tilde{A}\tilde{\xi}_k \tag{4.52}$$

where S is the solution of the Riccati equation

$$0 = \tilde{A}^\top S \tilde{A} - S + \tilde{Q} - \tilde{A}^\top S \tilde{B}(\tilde{R} + \tilde{B}^\top S \tilde{B})^{-1} \tilde{B}^\top S \tilde{A}. \tag{4.53}$$

The original control input is recovered by

$$u(k) = u(k-1) + \tilde{u}_k. \tag{4.54}$$

For our purpose, we choose the continuous-time longitudinal dynamics to be of the form

$$\dot{v} = a \tag{4.55}$$

with system output $y = v$ which fulfills Assumptions 26 and 27. By Euler-forward discretizing eq. (4.55) with step size h, we result in

$$\begin{aligned}
\xi(k+1) &= A\xi(k) + Bu(k), \quad \xi(0) = \xi_0 \\
y(k) &= C\xi(k),
\end{aligned} \tag{4.56}$$

with $\xi(k) = v(k)$, $u(k) = a(k)$, $A = 1$, $B = h$ and $C = 1$. Moreover, we choose the tuning matrices to be $Q = 1$ and $R = 0.1$.

4.4 Simulation results and experiments

In this section, we perform simulation tests of our algorithms proposed in Subsections 4.2.1, 4.2.2 and Section 4.3 before we show experimental results. As already mentioned in Section 4.1, we use the nonlinear kinetic single track model as a continuous-time plant model and the linearized kinematic single track model as the system model of the rbMPC algorithms defined in Subsections 4.2.1 and 4.2.2. In terms of the system model linearized along the trajectory as a i-th step ahead predictor, we have

$$A_i(k) = \begin{bmatrix} 1 & 0 & -h\,v_{\text{ref}}(k+i)\sin(\gamma_{\text{ref}}(k+i)) & h\cos(\gamma_{\text{ref}}(k+i)) \\ 0 & 1 & h\,v_{\text{ref}}(k+i)\cos(\gamma_{\text{ref}}(k+i)) & h\sin(\gamma_{\text{ref}}(k+i)) \\ 0 & 0 & 1 & \frac{h}{l_r}\sin(\beta_{\text{ref}}(k+i)) \\ 0 & 0 & 0 & 1 \end{bmatrix}, \tag{4.57}$$

$$B_i(k) = \begin{bmatrix} 0 & -h\,v_{\text{ref}}(k+i)\sin(\gamma_{\text{ref}}(k+i)) \\ 0 & h\,v_{\text{ref}}(k+i)\cos(\gamma_{\text{ref}}(k+i)) \\ 0 & h\,\frac{v_{\text{ref}}(k+i)}{l_r}\cos(\beta_{\text{ref}}(k+i)) \\ h & 0 \end{bmatrix} \tag{4.58}$$

with $\gamma_{\text{ref}}(k+i) = \psi_{\text{ref}}(k+i) + \beta_{\text{ref}}(k+i)$. The system state and the control input are defined as $\xi(k) = [x(k) \ y(k) \ \psi(k) \ v(k)]^\top \in \mathbb{R}^4$ and $u(k) = [\beta(k) \ a(k)]^\top \in \mathbb{R}^2$ at sampling instant $k \in \mathbb{N}$, respectively. The sampling time of the controller is chosen to be $T_c = 4 \cdot 10^{-2}$s. The parameters of the algorithms proposed in Subsections 4.2.1 - 4.2.2 and Section 4.3 are given below.

- *RbMPC trajectory tracking controller:* The tuning variables for the rbMPC trajectory controller are given below.

rbMPC trajectory tracking algorithm	Newton-based optimization algorithm
$N = 30$	$\epsilon_N = 1e^{-3}$
$Q = \text{diag}(0.9e^{-3}, 0.9e^{-3}, 3e^{-2}, 1e^{-1})$	$\alpha = 1e^{-2}$
$R = \text{diag}(1e^{-1}, 1e^0)$	$\beta = 0.9$
$P = Q$	
$\delta = 1e^{-7}$	
$\epsilon = 1e^{-5}$	

Table 4.1. Tuning variables for the rbMPC trajectory tracking algorithm and for the Newton-based optimization algorithm. For the rbMPC trajectory tracking algorithm we choose the prediction horizon N, the tuning matrices (Q, R), the terminal cost P, the relaxation parameter δ and the weighting parameter of the relaxed barrier function ϵ. For the Newton-based optimization algorithm, we select the decrease parameter α, the level of the line search's crudity β and the parameter that describes the level of optimality ϵ_N.

Moreover, the optimizer and the previous input are initialized with $u(-1) = 0 \in \mathbb{R}^m$ and $\Delta U(0) = 0 \in \mathbb{R}^{Nm}$. Besides, the number of Newton iterations $i_T(k)$ is restricted to be $i_T(k) \leq 100$. Furthermore, we choose the state constraints to be inactive.

- *RbMPC path following controller:* The tuning parameter of the rbMPC path following controller and its associated Newton-based optimization algorithm are chosen according to Table 4.1. Additionally, we impose $v_{min} = -0.99$, $v_{max} = 1$, $\theta^*(-1) = [1, 2, \ldots, N]^\top$ and $T = 100$.

- *Pure pursuit with LQR in velocity form:* The pure pursuit algorithm representing the lateral control has only one tuning parameter, the so-called look-ahead distance L_l. In the following, we will choose the look-ahead distance to be 8 points in front of the closest point of the actual vehicle's position toward the reference position trajectory. Hence, the look-ahead distance L_l is time-varying as the reference trajectory is time-varying. The longitudinal controller representing the LQR velocity controller consists of two tuning matrices, which are chosen to be $Q = 1$ and $R = 0.1$.

Moreover, we impose the following input and Δ input constraints

$$\begin{bmatrix} \beta_{min} \\ a_{min} \end{bmatrix} \leq u(k) \leq \begin{bmatrix} \beta_{max} \\ a_{max} \end{bmatrix}, \quad \begin{bmatrix} \Delta\beta_{min} \\ \Delta a_{min} \end{bmatrix} \leq \Delta u(k) \leq \begin{bmatrix} \Delta\beta_{max} \\ \Delta a_{max} \end{bmatrix} \quad (4.59)$$

with

β_{min}	β_{max}	a_{min}	a_{max}	$\Delta\beta_{min}$	$\Delta\beta_{max}$	Δa_{min}	Δa_{max}
-0.3531rad	0.3531rad	$-15\frac{m}{s^2}$	$7\frac{m}{s^2}$	-0.046rad	0.046rad	$-3\frac{m}{s^2}$	$3\frac{m}{s^2}$.

Table 4.2. Parameters of the imposed constraints

Note that β_{min}, β_{max} and $\Delta\beta_{min}, \Delta\beta_{max}$ refers via $\beta = \text{atan}(\frac{l_r}{l_f+l_r})\tan(\delta_f\frac{\pi}{180}))$ to $\delta_{min} = -35°, \delta_{max} = 35°$ and $\Delta\delta_{f,min} = -5°, \Delta\delta_{f,max} = 5°$. The corresponding polytopic input constraints $C_u \leq d_u$ and Δ input constraints $C_{\Delta u}\Delta u \leq d_{\Delta u}$ are defined as

$$\begin{aligned} C_u &= \begin{bmatrix} 1 & 0 \\ 0 & 1 \\ -1 & 0 \\ 0 & -1 \end{bmatrix}, \quad d_u = \begin{bmatrix} a_{max} \\ \delta_{f,max} \\ -a_{min} \\ -\delta_{f,min} \end{bmatrix}, \\ C_{\Delta u} &= \begin{bmatrix} 1 & 0 \\ 0 & 1 \\ -1 & 0 \\ 0 & -1 \end{bmatrix}, \quad d_{\Delta u} = \begin{bmatrix} \Delta a_{max} \\ \Delta\delta_{f,max} \\ -\Delta a_{min} \\ -\Delta\delta_{f,min} \end{bmatrix}. \end{aligned} \quad (4.60)$$

In the following, we will show simulation results of the three algorithms (1), (2) and (3) at the different maneuvers and different performance criteria with the tuning variables and constraints as defined above.

4.4.1 Constant circular drive ISO 4138

As a first test maneuver, we choose the circular test-drive in anti-clockwise direction with constant radius $r = 100$m and constant velocity $v_{\text{ref}} = \frac{70}{3.6} \frac{\text{m}}{\text{s}}$, according to ISO 4138, see M. Bargende et al. (2019)[Section 3.2], in order to check the steady-state behavior and the ability to deal with big initial offsets of the proposed controllers. For the sake of clarity, we only show the first seconds of the maneuver because it already contains the crucial tracking behavior.

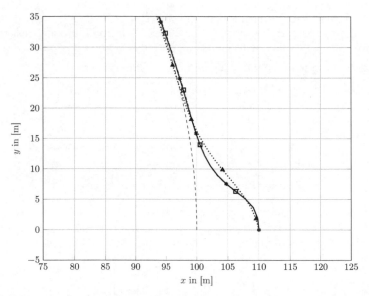

Figure 4.10. Closed-loop behavior of the position coordinates (x, y) of the constant circular drive ISO 4138 with a constant radius $r = 100$m and a constant speed of $v = \frac{70}{3.6} \frac{\text{m}}{\text{s}}$. The figure depicts the reference position (---), the closed-loop behavior of the rbMPC tracking (—✦—), the rbMPC path following (-ᴇ-) with a maximal Newton iteration number of $i_{\text{T}}(k) \leq 100$ steps and the pure pursuit control algorithm (··▲··).

From Figure 4.10, we can conclude that the rbMPC tracking, the rbMPC path following and the pure pursuit algorithm drive the system state towards the reference state trajectory. Moreover, the rbMPC algorithms have the same driving line, whereas the pure pursuit control algorithm possesses a different transient behavior. This effect originates from the different structure of the pure pursuit algorithm and can be affected by the choice of the look-ahead distance L_l.

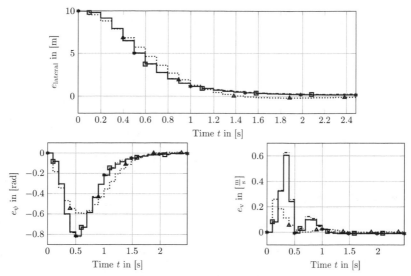

Figure 4.11. Evolution of the lateral error e_{lateral}, the heading error e_ψ, and the velocity error e_v for the constant circular drive ISO 4138 with constant radius $r = 100$m and constant speed $v = \frac{70}{3.6} \frac{\text{m}}{\text{s}}$ of the rbMPC tracking ($\rightarrow$), the rbMPC path following (-$\boxminus$-) and the pure pursuit control algorithm ($\cdots\triangle\cdots$).

From Figure 4.11, we can conclude that the transient behavior of the lateral error of the rbMPC tracking, rbMPC path following, and pure pursuit algorithm is similar. The actual system state approaches to the reference system state without overshooting. The ability of time-reparametrization of the path following rbMPC algorithm compared with the trajectory tracking rbMPC algorithm does not show any advantage with respect to improved tracking behaviour. Moreover, we can conclude that the pure pursuit algorithm performs slightly better in the yaw error and velocity error, which is an effect of the chosen look-ahead distance L_1. Furthermore, the rbMPC tracking and the rbMPC path following algorithm result in a small steady-state offset. The steady-state offset is caused by the rbMPC formulation which can be compensated in future by either using a disturbance observer or reformulating the rbMPC algorithms in full velocity form, similar to the velocity controller of the pure pursuit control algorithm, see D. Ruscio (2013); G. Pannocchia, M. Gabiccini, A. Artoni (2015); M. Stephens and M. Good (2013). Nevertheless, we will keep the current rbMPC formulations as the lateral steady-state error size is small enough. Moreover, we will mark the so-called offset-free MPC formulation as a further extension.

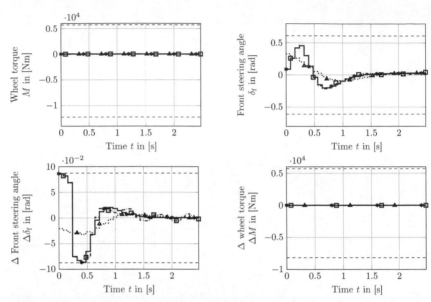

Figure 4.12. Evolution of the control inputs M and δ_f and the Δ control inputs ΔM and $\Delta\delta_f$ of the constant circular drive ISO 4138 with constant radius $r = 100$m and constant speed $v = \frac{70}{3.6}\frac{m}{s}$. The figure depicts the rbMPC tracking $(-\!\!\ast\!\!-)$, the path following rbMPC $(\cdot\text{-}\square\text{-}\cdot)$ and the pure pursuit control algorithm $(\cdots\!\scriptstyle\blacktriangle\cdots)$. The imposed constraints are depicted with $(\text{-}\text{-}\text{-})$.

In almost any practical application, a dynamical system is restricted to physical constraints. In our case physical constraints are modeled as system state or control input constraints. In general, the control goal is not only to guarantee tracking toward the reference but also the strict satisfaction of the system state or control input constraints. In case of the rbMPC algorithms (1) and (2), the constraints are explicitly taken into account through the use of relaxed barrier functions, see Section 1.3. Depending on the initial state and the choice of the relaxation parameter δ, exact constraint satisfaction for a certain set of initial conditions can always be guaranteed, see Chapter 2. Due to the circular drive carried out with constant velocity (and the initial condition being equal to that velocity), we see from Figure 4.12 that the actual wheel and Δ wheel torque are near to zero. Thus, the actual values stay away from their respective bounds. However, owing to the large initial position error, we see that the value of the steering wheel and Δ steering wheel signal evolves close to or lies exactly on its respective bounds. Therefore, we can conclude that the rbMPC tracking algorithm and the rbMPC path following algorithm strictly satisfy the imposed constraints. To guarantee strict constraint satisfaction for the pure pursuit algorithm, we simply restrict the control input and Δ control input with suitable parametrized saturation blocks. From Figure 4.12, we can conclude that for the pure pursuit algorihtm, the actual front steering angle and Δ front steering angle are as well satisfied.

Similar to the rbMPC algorithms, the actual wheel torque and Δ wheel torque values stay away from their respective bounds, due to the definition of the maneuver. In the following, we will examine the computational effort needed to calculate the control input $u(k)$ for the rbMPC tracking, the rbMPC path following, and the pure pursuit algorithm.

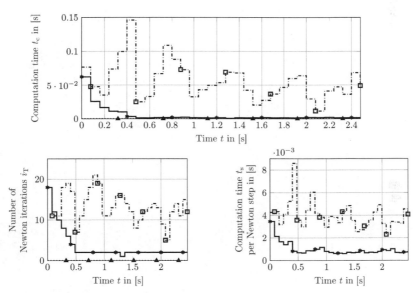

Figure 4.13. Evolution of the absolute computation time t_c, the computation time of one Newton step t_s, and the overall performed Newton steps $i_T(k)$ for the constant circular drive ISO 4138 with constant radius $r = 100$m and constant speed $v = \frac{70}{3.6}\frac{m}{s}$. The figure depicts the rbMPC tracking ($-\!\!*\!\!-$), the path following rbMPC ($-\square-$) algorithm, and the pure pursuit algorithm ($\cdots\!\triangle\!\cdots$).

As the rbMPC tracking and the rbMPC path following algorithm perform an online optimization and are implemented on a real-time target machine that possesses only finite computation power, it is mandatory that the overall computation time t_c stays below the sampling time of the controller T_c. Figure 4.13 shows that the rbMPC path following algorithm has a huge overall computation time of about 0.15s at its maximum, corresponding to 20 performed Newton steps. Thus, it will be difficult to ensure that $t_c \leq T_c$ on a real-time target machine at a real vehicle. Nevertheless, we know from C. Feller (2017) that the reformulation with relaxed barrier functions possesses the so-called anytime property that enables us to stop the Newton iteration after, for example, one step and apply the suboptimal control input to the system. With a single Newton step lasting at its maximum $5 \cdot 10^{-3}$s, it would be possible to ensure that $t_c \leq T_c$.

For the performance differences on the application of suboptimal inputs to the closed loop, we refer the reader to F. Pfitz, C. Ebenbauer and M. Braun (2019). Moreover, we want to mention that the rbMPC tracking algorithm possesses an averaged overall computation time of about 3ms, making it attractive for the implementation on a real-time target device in the real vehicle. Furthermore, the pure pursuit controller performs best with respect to the overall computation time (almost zero, and thus not visible in the figure). This fact is obvious, because it possesses rather simple calculations compared to the rbMPC tracking and the rbMPC path following algorithm.

4.4.2 Dual lane change ISO 3888-2

The dual lane change ISO 3888-2 was originally designed to evaluate the handling performance of a vehicle, see Vehico GmbH (2020) with a total length of 61m. We propose a target velocity of $12\frac{m}{s}$. It is characterized by an entry and exit lane length of 12m and a side lane length of 11m. The width of the entry, side lane, and exit lane are defined to be 3m. The lateral offset between the side and exit lane as well as between the side and entry lane is 1m. The longitudinal offset between the entry and the side lane amounts to 13.5m, whereas the longitudinal offset between the exit and the side lane amounts to 12.5m.

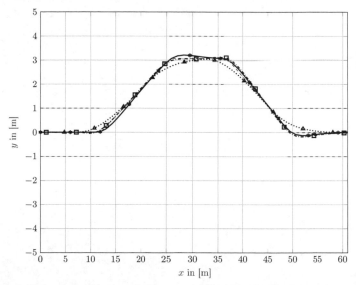

Figure 4.14. Closed-loop behavior of the position coordinates (x, y) of the dual lane change ISO 3888-2 with a constant velocity of $v = 12\frac{m}{s}$. The figure depicts the reference position and the allowed area (---). Moreover, it shows the closed-loop behavior of the rbMPC tracking (—*—), the rbMPC path following (-□-) with a maximal Newton iteration number of $i_T(k) \leq 100$ steps, and the pure pursuit control algorithm (··▲··).

From Figure 4.14, we can conclude that the proposed rbMPC tracking, the rbMPC path following, and the pure pursuit control algorithm stay within the dashed area at the entry, the side, and the exit lane. Thus, the rbMPC tracking, the rbMPC path following, and the pure pursuit algorithm pass the dual lane-change test. However, if we look at the x-coordinate between 10m and 20m, which represents the area around the entry lane, we see that the rbMPC path following algorithm uses its ability of time-reparametrization, see Section 2, to get better tracking performance compared to the trajectory tracking rbMPC algorithm. In the case of the pure pursuit algorithm, we see that the reference is tracked predictively, because of the choice of the look-ahead distance L_l. Besides, in case of the rbMPC tracking and path following algorithm, the transient behavior depends greatly on the choice of the internal prediction model. Note that we used the kinematic single track model as internal prediction model and the kinetic single track model as a simulation model which leads to an overshoot at the exit lane for both the rbMPC tracking and the rbMPC path following algorithm.

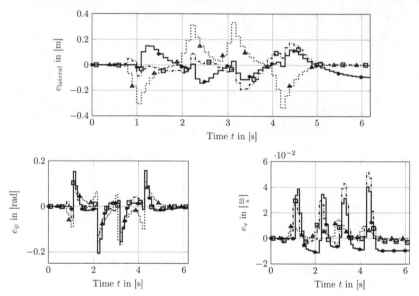

Figure 4.15. Evolution of the lateral error e_{lateral}, the heading error e_ψ and the velocity error e_v of dual lane change ISO 3888-2 with constant velocity $v = 12\frac{m}{s}$. The figure depicts the rbMPC tracking (—∗—), the rbMPC path following (-□-), and the pure pursuit algorithm (··▲··).

From Figure 4.14, we can conclude that the pure pursuit algorithm is the algorithm that follows most smoothly the reference trajectory. Figure 4.15 confirms this observation, showing the lowest error in heading e_ψ and velocity e_v.

However, if we look at the maximal lateral error, we see that the pure pursuit algorithm performs worst at about 25cm. The rbMPC tracking and path following algorithm perform better with a maximal lateral error of less than 20cm. If we look at the entry lane, which is represented by the time between 0.8s and 1.5s, we see that the path following rbMPC algorithm performs much better than the corresponding trajectory tracking rbMPC algorithm. This effect can be explained by its additional capacity of time re-parametrization, see Chapter 2.

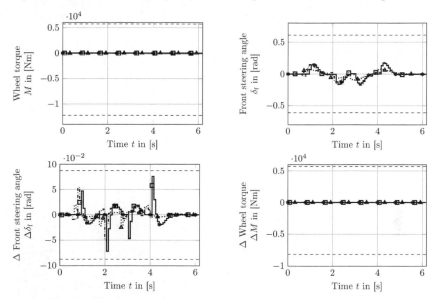

Figure 4.16. Evolution of the control inputs M, δ_f, and the Δ control inputs ΔM, $\Delta \delta_f$ of the dual lane change ISO 3888-2 with constant velocity $v = \frac{m}{s}$. The figure depicts the rbMPC tracking ($\rightarrow$), the path following (-□-) MPC algorithm, and the pure pursuit algorithm ($\cdots\triangle\cdots$). The imposed constraints are depicted in (---).

As already mentioned in the previous maneuver, in Subsection 4.4.1, we know that any practical application has to satisfy physical constraints. Thus, not only the tracking behavior plays an important role, but also the satisfaction of the physical constraints. From Figure 4.16, we can conclude that the input constraints, as well as the Δ input constraints, are strictly satisfied for rbMPC tracking, the rbMPC path following, and the pure pursuit algorithm. As the pure pursuit algorithm follows most smoothly the reference, it stays most far away from the corresponding constraints. In the following, we will continue with the evaluations of the computation time which is the time needed for the calculation of the control input $u(k)$ for the rbMPC tracking, the rbMPC path following, and pure pursuit algorithm.

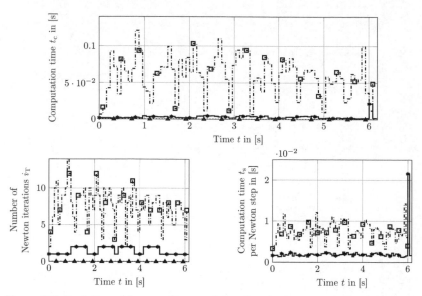

Figure 4.17. Evolution of the overall computation time t_c, the computation time of one Newton step t_s, and the overall performed Newton steps $i_T(k)$ for the dual lane change ISO 3888-2 with constant velocity $v = 12\frac{m}{s}$. The figure depicts the rbMPC tracking (—∗—), the path following rbMPC algorithm (-□-) , and the the pure pursit algorithm algorithm (··▲··).

As for the previous maneuver of a constant circular drive, see Subsection 4.4.1, we can conclude that the pure pursuit algorithm involves rather simple calculations, compared with the rbMPC algorithms. Thus, it has the lowest overall computation time (near to zero). Furthermore, we see again that the rbMPC path following algorithm consumes the most computational power by having a maximal overall computation time of about 0.12s, corresponding to 18 Newton iterations. The rbMPC trajectory tracking algorithm performs better with an average of two Newton iterations and an average computation time of about 2ms. In terms of the application to a real-time target machine possessing only finite computation time, we like to mention that the rbMPC trajectory tracking and the pure pursuit algorithm are most realistic to implement to ensure that $t_c \leq T_c$.

4.4.3 Racetrack

To check the performance of the proposed algorithms in Subsections 4.2.1, 4.2.2 and Section 4.3 in terms of a lateral and longitudinal dynamic reference trajectory, we make use of a recorded endurance testing drive on the test facility in Weissach (Porsche AG, Germany). A graphic illustration of the test facility, the position coordinates of the endurance testing, and its corresponding velocity profile are given below.

Figure 4.18. Test facility in Weissach, Germany, see Porsche (2020).

The endurance testing profile was originally designed to test each component on its durability by longitudinally and laterally exciting the vehicle.

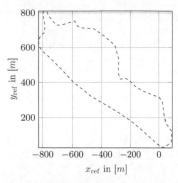
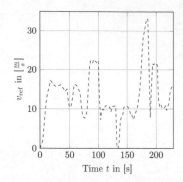

Figure 4.19. Position coordinates of the endurance testing profile in Weissach, Germany and its corresponding velocity profile.

Including two anti-blocking-system (ABS) brakes and sharp turns, the endurance testing profile is a demanding reference trajectory to deal with. From measurements, we can extract the minimal and maximal sideslip angle and Δ sideslip angle $\beta_{\min} = -0.0466$rad, $\beta_{\max} = 0.0749$rad, $\Delta\beta_{\min} = -0.0026$rad, and $\Delta\beta_{\max} = 0.0031$rad as well as the minimal and maximal accelerations and Δ accelerations $a_{\max} = 8\frac{m}{s^2}$, $a_{\min} = -11\frac{m}{s^2}$, $\Delta a_{\max} = 1.5\frac{m}{s^2}$ and $\Delta a_{\min} = -1.5\frac{m}{s^2}$. These values are valid as can be seen for comparable vehicles, see H. Burg and A. Moser (2017)[p. 470 Table A12.15]. In the following, we will continue with evaluations with respect to the tracking behavior of the rbMPC tracking, the rbMPC path following and the pure pursuit algorithm. For the sake of clarity, we only show the maneuver from $60s$ - $140s$ because it contains the most deviation in the tracking errors.

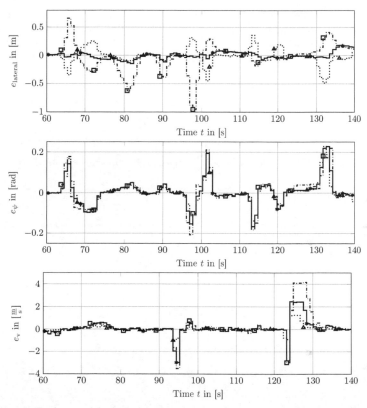

Figure 4.20. Evolution of the lateral error e_{lateral}, the heading error e_ψ and the velocity error e_v of the PG endurance testing drive with maximal velocity of $v = \frac{120}{3.6}\frac{\text{m}}{\text{s}}$. The figure depicts the rbMPC tracking (—•—), the rbMPC path following (-□-), and the pure pursuit control algorithm (··▲··).

From Figure 4.20, we can conclude that the rbMPC trajectory tracking and pure pursuit algorithm stay almost always lower than a 20cm lateral error during the whole PG endurance testing profile. The path following rbMPC controller performs worst, with roughly 90cm maximum of lateral deviation, which can be seen as a tuning issue. Moreover, the rbMPC path following algorithm possesses the biggest deviation in velocity of around $-4\frac{m}{s}$ at time $t = 130$s. In the latter, we will neglect the figures for the control input and Δ control input because they are less informative. We want to mention that with the choice of the relaxation parameter δ and the weighting parameter ϵ of the relaxed barrier functions, the inputs, and the Δ inputs, constraints are strictly satisfied over the whole endurance testing drive.

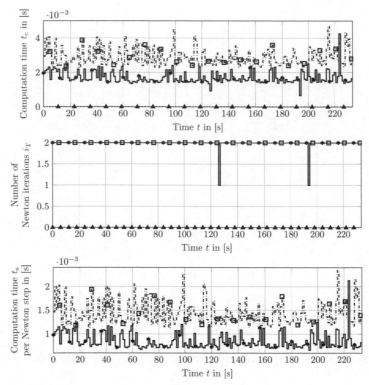

Figure 4.21. Evolution of the overall computation time t_c, the computation time of one Newton step t_s, and the overall performed Newton steps $i_T(k)$ for the racetrack drive associated with the endurance testing profile. The figure depicts the rbMPC tracking ($-\!*\!-$), the path following ($-\square-$) rbMPC algorithm, and the pure pursuit algorithm ($\cdots\!\triangle\!\cdots$).

As in the previous maneuvers of a constant cicular drive and a dual lane change, not only the tracking behavior from Figure 4.20 plays an important role for the practical application, but also the ensurance that the computation time t_c stays below the controller sampling time T_c.

102

Again, we have chosen the sampling time of the controller to be $T_c = 40\text{ms}$. With a maximal computation time of the rbMPC tracking, the rbMPC path following and the pure pursuit algorithm being less than 4ms, we can conclude that $t_c \leq T_c$, which means that the timing constraint is not violated and all algorithms can be implemented on a real-time target machine. However, the rbMPC tracking algorithm possesses the best tracking behavior, see Figure 4.20. From a theoretical point of view, this should not be the case as the path following algorithm posses an additional degree of freedom to adapt towards the reference. Thus, we can conclude that the simplifications of the approximated algorithm described in 4.2.2 have a negative effect on the overall performance of the algorithm for the racetrack drive. This has to be further investigated and is a topic for further research. In the latter, we recommend implementing the rbMPC tracking algorithm as it has a moderate calculation time and the best tracking results over the three different maneuvers. In the latter, we will implement the rbMPC tracking algorithm at a real vehicle of a Panamera G2 turbo.

4.4.4 Hardware configuration

Before we show the experimental results of an automated driving vehicle, we first describe the used hardware and the testing procedure of an automated driving vehicle. The testing vehicle is a V8 Panamera G2 turbo (404kW) equipped with rear-wheel steering, which we neglect in our modeling, see Section 4.1.

Figure 4.22. Testing vehicle Panamera G2 turbo.

Moreover, the Panamera G2 turbo is modified in several ways to be able to drive autonomously.

- *Sensors:* To read in the current system state, the Panamera G2 Turbo is equipped with an ADMA-G-Pro+ from Genesys that uses a combination of the GPS signal and the correction data of a base station together with accelerometers and gyroscopes of an inertial measurement unit (IMU) to achieve an absolute tolerance in position of ±2cm with a sampling time of 2ms.

- *Real-time target machine:* The trajectory tracking rbMPC algorithm from Chapter 2 is developed in Matlab/Simulink and is loaded via Simulink Realtime based on automatic code compilation on a Speedgoat mobile target (Intel i5-7300 vPro processor 12.51 GFLOPS). The Speedgoat is connected to the ADMA-G-Pro+ to get the current system state. The reference trajectory is stored offline and is used together with the current system state to calculate the control input.

- *Actuators:* Via low-level controllers, the control input is transformed and applied to the vehicle interfaces. Depending on the signal itself, each signal is transferred over CAN/Flexray with a sampling time of 10ms up to 50ms. In the Panamera G2 Turbo, we use the manipulation of the throttle pedal to accelerate, the manipulation of an electronic brake booster (EBS) to decelerate, and the manipulation of the electronic power steering system (EPS) to steer.

A successful test run consists of the following steps

Step 1: Place the vehicle at some starting point $(x_{ref}(0), y_{ref}(0))$ in the inertial frame.

Step 2: Set the vehicle in record mode and record the reference trajectory (manual drive) with the ADMA-G-Pro+, and store the reference trajectory offline.

Step 3: Place the vehicle in the neighborhood of the starting point $(x_{ref}(0), y_{ref}(0))$ (see Step 1).

Step 4: Set the vehicle in autonomous mode, release the brake, and start the maneuver.

Step 5: The vehicle stops at the end of the trajectory and the driver takes back control of the vehicle.

Figure 4.23. Steps of a successful automated test run.

4.4.5 Experimental Results

Based on the hardware configuration described in the previous section, we now implement the rbMPC tracking algorithm from Subsection 4.2.1. The implementation and data aggregation was mainly carried out by T. Schanz (2018). However, we will revise the most important facts to show the results as a proof of concept to the reader. The reference position is characterized by two 270°-turns with a radius of about 20m in the opposite direction and a lane change in between. The maximal velocity is restricted to $9\frac{m}{s}$ for security reasons. The measured maximal longitudinal and lateral acceleration was $2.7\frac{m}{s^2}$ and $5.6\frac{m}{s^2}$, respectively. The test run follows along the steps of (1). The sampling time of the controller is chosen to be $T_c = 60$ms. Furthermore, we impose the following control input and Δ control input constraints

$$\begin{bmatrix} \beta_{min} \\ a_{min} \end{bmatrix} \leq u(k) \leq \begin{bmatrix} \beta_{max} \\ a_{max} \end{bmatrix}, \tag{4.61}$$

$$\begin{bmatrix} \Delta\beta_{min} \\ \Delta a_{min} \end{bmatrix} \leq \Delta u(k) \leq \begin{bmatrix} \Delta\beta_{max} \\ \Delta a_{max} \end{bmatrix}. \tag{4.62}$$

with

β_{min}	β_{max}	a_{min}	a_{max}	$\Delta\beta_{min}$	$\Delta\beta_{max}$	Δa_{min}	Δa_{max}
-0.3531rad	0.3531rad	$-2\frac{m}{s^2}$	$1\frac{m}{s^2}$	-2.1rad	2.1rad	$-0.1\frac{m}{s^2}$	$0.1\frac{m}{s^2}$.

Table 4.3. Parameters of the imposed constraints

Note that β_{min}, β_{max}, and $\Delta\beta_{min}, \Delta\beta_{max}$ refers via $\beta = \text{atan}(\frac{l_r}{l_f+l_r})\tan(\delta_f\frac{\pi}{180}))$ to $\delta_{min} = -35°, \delta_{max} = 35°$ and $\Delta\delta_{f,min} = -T_c\frac{35°}{s}, \Delta\delta_{f,max} = T_c\frac{35°}{s}$. The corresponding polytopic input and Δ input constraints are obtained similar to (4.60). The relaxation parameter and weighting parameter are $\delta = 1e^{-6}$ and $\epsilon = 1e^{-2}$ to ensure strict constraint satisfaction. The initialization of the previous control input, the optimizer, and the parameter of Newton's method are selected according to Section 4.2. Moreover, we choose the tuning matrices (Q, R) and the terminal cost P to

$$Q = \text{diag}(0.1, 0.1, 120, 30), \quad R = \text{diag}(10, 1e^3), \quad P = Q. \tag{4.63}$$

An illustration of the described test run is given in Figure 4.24.

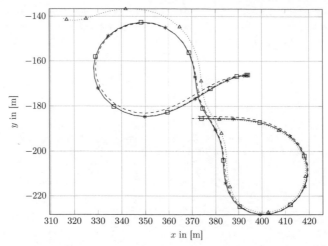

Figure 4.24. Experimentally measured closed-loop behavior for the described maneuver with a maximal speed of $v = 9\frac{m}{s}$ in the inertial frame (X, Y) at position (x, y). The figure depicts the reference trajectory (---) , the resulting closed-loop trajectory fully iterated (-□-) , the closed-loop trajectory fully iterated with a big initial offset (··△··) , and the closed-loop trajectory with a single Newton iteration $i_T(k) = 1$ (-*-) .

Figure 4.24 shows the performance of the proposed rbMPC trajectory tracking algorithm. All closed-loop trajectories are converging toward the reference trajectory, the proposed rbMPC algorithm is even able to cope with large initial errors (··△··) . In the latter, we will restrict ourselves to the case of small initial errors with full iteration (-□-) and a single Newton step (-*-) . Moreover, it is important to note that the following results were gathered without the actuation of the electronic brake booster, because of an internal error of the braking system. Hence, the vehicle does not decelerate at the end of the trajectory and we truncate the measured signals and the reference trajectories when the vehicle starts to brake.

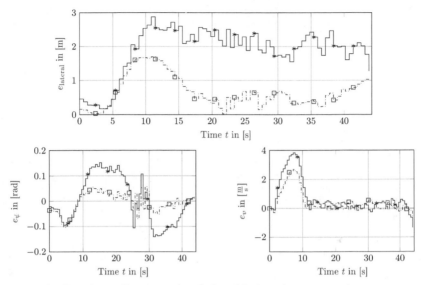

Figure 4.25. Experimentally measured evolution of the lateral error e_{lateral}, the yaw error e_ψ and the velocity error e_v of the described maneuver, see Figure 4.24.

Figure 4.25 shows the tracking performance of the system state with small initial error at full iteration (-◻-) and at a single Newton iteration (—✱—) . For the sake of clarity, we neglect the test run with big initial error (⋯▵⋯) in the evaluations because it badly scales the lateral error figure. For the qualitative result, we refer the reader to Figure 4.24, corresponding to the trajectory (⋯⋯). Although, we have huge deviations in the lateral error of maximal 1.6m in case of full iteration (-◻-) and 2.5m in case of a single Newton step (——), we see that the full iteration test run (-◻-) has less lateral error than the test run, corresponding to a single Newton iteration. Note that the shuttering of the signals, as well as the huge deviations in the lateral error, originate from badly tuned low-level controllers and huge dead-times in the actuators, also evaluated by T. Schanz (2018). To sum up, we have implemented the rbMPC trajectory tracking at the real vehicle. Due to insufficient precise actuators, it was not possible to test the performance of the controller at a satisfactory level. A possible solution to the problem would be to model the dynamics of the actuators (including the dead times) directly in the MPC internal prediction model to achieve improved tracking performance. However, we have seen that the proposed rbMPC steers stabilizes the vehicle around the desired trajectory, see Figure 4.24, which results in bounded errors, see Figure 4.25, which can be seen as a proof of concept.

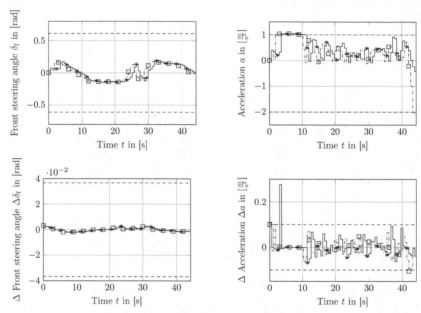

Figure 4.26. Experimentally measured evolution of the control input $u = \delta_f$ and the Δ control input $\Delta u = \Delta \delta_f$ of the described maneuver, see Figure 4.24.

As mentioned in the simulation tests in Subsection 4.4.1 - 4.4.3, almost any practical application is restricted to physical constraints such as steering angle constraints. Depending on the physical constraint itself, we have to guarantee strict constraint satisfaction in order to avoid severe damage of the actuator. For example, think of steering angle and Δ steering angle constraints. A constraint violation of the steering angle constraint may lead to severe damage of the steering wheel, whereas a violation of the Δ steering angle constraints does not cause any harm to the actuator. Figure 4.26 evaluates the control input and Δ control input constraints. From Figure 4.26, we can conclude that the absolute value of the control inputs stay far away from the actual constraint limit. Hence, the actuator is not damaged. In case of the Δ control inputs, we see in Figure 4.26 that it is almost satisfied for full iteration (-⊟-) and violated for a single Newton step at the Δ acceleration at time $t = 3s$ which is not harmful to the actuator. Note that the open-loop optimal control problem of the rbMPC trajectory tracking algorithm is still feasible although lying outside of the actual constraint set, due to the formulation of the constraints by relaxed barrier functions.

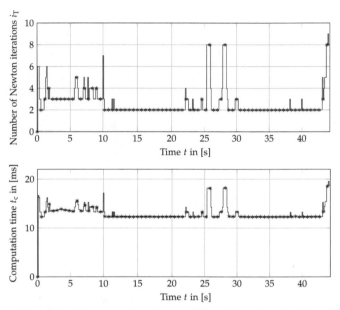

Figure 4.27. Experimentally measured evolution of the total number of performed Newton steps in case of full iteration (-∗-) and the overall computation time t_c of the proposed algorithm.

Because of MPC being an online optimization scheme, it is mandatory to check if the overall computation time t_c stays below the sampling time of the controller T_c. We have imposed a controller sampling time of $T_c = 60ms$, based on Figure 4.27, we can conclude that the computation time t_c is always less than $T_c = 60ms$ and thus satisfies the timing constraint. Moreover, Figure 4.27 shows the total number of performed Newton iterations $i_T(k)$ as well as the overall computation time. We see that only two Newton iterations are needed to compute the optimum around $12.5 - 22.5s$ and $30 - 38s$, which correspond to the two sharp $270°$ turns where the velocity and yaw rate are nearly constant. Owing to the Δ control inputs being nearly exploited between $0 - 10$s more Newton iterations $i_T(k)$ are needed to compute the optimum.

In this chapter, we modified the rbMPC tracking and path following algorithm from Chapter 2 such that it can be applied to nonlinear systems. Furthermore, we like to mention that we introduced a Δ input weighting into the cost functionality. Moreover, we gave an overview of different dynamical models which are either used as internal prediction of the proposed rbMPC algorithms or as simulation model for the closed loop. Furthermore, we developped a state of the art benchmark controller which consists of a decoupled lateral and longitudinal controller. We defined three difference reference maneuvers: 1.) Constant circular drive ISO 4138, 2.) Dual lane change ISO 3888-2 and 3.) a racetrack drive and we compared the modified rbMPC tracking and path following algorithms as well as the benchmark controller according to the reference maneuvers with respect to tracking performance, constraint satisfaction guarantees and computational complexity analysis. Finally, we point out that the rbMPC tracking algorithm is the best performing algorithm with respect to tracking behavior and computational affordability. Hence, we implement the rbMPC trajectory tracking algorithm at the real vehicle. We provide real data to the reader in terms of tracking performance and computational complexity and give advice for further improvement.

5
Summary and conclusion

In the following, we will give a summary of the thesis and highlight areas for further research. Above all, we will discuss the most important findings from a conceptual, numerical, and practical point of view.

Summary and Conclusion: The key idea and main motivation of this thesis was the investigation of new algorithms of a driving dynamics controller that could cope with demanding driving tasks, such as driving on a racetrack and dealing with low computational power. Basically the developed algorithms can be classified as *coupled* and *decoupled* algorithms which differ in its longitudinal and lateral driving dynamics interconnection. As the term *coupled* suggests, a coupled controller considers the coupling between the longitudinal and lateral dynamics, while a decoupled controller considers the longitudinal and lateral dynamics to be separated. In this thesis, we have focused on coupled predictive control algorithms, which naturally possesses the ability to take constraints into account. Basically, there are two problem formulations that solve the problem task of following a given driving line. On the one hand, there is the so-called *trajectory* or *reference tracking problem*, and on the other hand, there is the *path following problem* which is more suitable to the autonomous driving problem, however it is more complicated. We investigated both problems based on a rbMPC formulation, see C. Feller (2017). In Chapter 2, we gave important detailed stability and constraint satisfaction guarantees of the trajectory tracking and path following problem based on rbMPC for linear discrete-time systems. Thus, we have extended the theoretical framework of rbMPC for the trajectory tracking and path following problem. Moreover, we evaluated in Sections 4.1 - 4.4 the proposed algorithms in simulation and provided experimental data in the area of an automated driving vehicle. From the simulations and experiments carried out in Section 4.4, we can conclude that the *trajectory tracking* is the most promising predictive control algorithm with respect to tracking a given driving line, computational efficiency, and constraint satisfaction. For a further extension of the proposed algorithms, we refer to F. Pfitz and M. Schaefer (2022). To overcome the loss of important system theoretical guarantees like stability, Chapter 3 is intended to design a learning-based predictive scheme. According to the *certainty equivalence principle*, an estimation algorithm and a receding horizon control algorithm were designed. The estimation algorithm could be thought of as an asymptotic correct signal predictor that does *not* perform any *model identification*, standing in contrast to most existing estimation schemes. Moreover, the predictive control algorithm was designed such that convergence to the origin of the closed-loop system is guaranteed purely depending on a terminal cost. It does not use any knowledge of the system parameters and hence is referred to as a *model-independent* receding horizon control scheme. Finally, the predictive control algorithm and the estimation algorithm were combined and convergence guarantees under minimal assumptions were carried out for linear discrete-time systems. Simulation results for both linear and nonlinear systems were provided, which show a possibly promising approach for nonlinear systems.

We conclude the thesis by pointing out possible areas of further research based on the made evaluations.

Future research: In Sections 4.1 - 4.4, we discussed simulation and experimental results of the proposed trajectory tracking and path following, see Chapter 2, rbMPC algorithms. We formulated the convex open-loop optimal control problem in a vector-matrix formulation and highlighted its numerical solution. The open-loop optimal control problem is formulated such that the desired system state is tracked asymptotically in the well-known Δu formulation. However, if the system dynamics suffer from constant disturbances, we result in a steady-state error. This effect is undesirable and can be solved by formulating the MPC scheme in an offset-free output space formulation, see G. Betti, M. Farina and R. Scattolini (2012); G. Pannocchia, M. Gabiccini, A. Artoni (2015); U. Maeder, F. Borrelli and M. Morari (2009). Hence, we propose the following further improvement. To guarantee an offset-free formulation, the MPC scheme has to be formulated with a suitable disturbance observer or the MPC scheme has to be in full velocity form, see D. Ruscio (2013). Therefore, it is mandatory that the internal prediction model is formulated in lateral error coordinates. In Chapter 3, we designed a learning-based predictive control scheme for the control of unknown and unconstrained systems. For future research, investigations of *constrained* systems can be carried out. Moreover, numerical aspects such as a *preinitialization* of the estimation vector as well as the application of suitable upper and lower *bounds* to the estimation vector can be considered, see e.g. N.

Saraf and A. Bemporad (2017); P. Bouffard, A. Aswani and C. Tomlin (2012). Furthermore, the stability analysis for *nonlinear* systems is missing. Finally, we refer to Assumption 15 c.) in Chapter 3 which is the most technical assumption in the overall learning based setup for the control of fully unknown systems. According to Remark 21, the estimation scheme has to be further investigated such that stabilizability of the estimated limit system can be guaranteed.

A
Stability of discrete-time systems

In the following , we will revise important stability concepts for discrete-time systems. In particular, we will focus on Lyapunov stability and input-to-state (ISS) stability concepts.

A.1 Lyapunov stability

We like to mention that most of the content is from standard literature and can be found for example in J. Rawlings, D. Mayne and M. Diehl (2019).

Lemma 10. If there exists a continuous differentiable function $V : \mathcal{D} \times \mathbb{N} \to \mathbb{R}_+$ and constants $a, b, c, \sigma > 0$ such that for all $x(k) \in \mathbb{R}^n$, it holds that

$$
\begin{aligned}
a||x(k)||^{\sigma} &\leq V(x(k), k) \leq b||x(k)||^{\sigma}, \\
V(f(x(k), k), k+1) - V(x(k), k) &\leq -c||x(k)||^{\sigma},
\end{aligned}
\tag{A.1}
$$

then the zero steady-state solution of $x(k+1) = f(x(k), k)$ is globally uniformly exponentially stable, where $x(k) \in \mathcal{D}$, $\mathcal{D} \subseteq \mathbb{R}^n$ such that $0 \in \mathcal{D}$, $f : \mathcal{D} \times \mathbb{N} \to \mathbb{R}^n$ is such that $f(\cdot, \cdot)$ is jointly continuous in k and x, and for every $k \in \mathbb{N}$, $f(k, 0) = 0$ and $f(k, \cdot)$ is locally Lipschitz in x uniformly in k for all $k \in \mathbb{N}$, see J. Rawlings, D. Mayne and M. Diehl (2019)[Section 2.4, Definition 2.31] and W. Haddad and V. Chellaboina (2011)[Chapter 4, Theorem 4.6ii] •

Lemma 11. Suppose there exists a continuous differentiable function $V : \mathbb{R}^{n_1} \times \mathbb{R}^{n_2} \to \mathbb{R}_+$ and positive constants $a, b, c, \sigma \geq 1$ that satisfy

$$
\begin{aligned}
a||x_1(k)||^{\sigma} &\leq V(x_1(k), x_2(k)) \leq b||x_1(k)||^{\sigma}, \\
V(f_1(x_1(k), x_2(k)), f_2(x_1(k), x_2(k))) - V(x_1(k), x_2(k)) &\leq -c||x_1(k)||^{p}
\end{aligned}
\tag{A.2}
$$

for the nonlinear discrete-time system

$$
\begin{aligned}
x_1(k+1) &= f_1(x_1(k), x_2(k)), \quad x_1(0) = x_{10}, \\
x_2(k+1) &= f_2(x_1(k), x_2(k)), \quad x_2(0) = x_{20},
\end{aligned}
\tag{A.3}
\tag{A.4}
$$

where $x_1 \in \mathcal{D}$, $\mathcal{D} \subseteq \mathbb{R}^{n_1}$ is an open set such that $0 \in \mathcal{D}$, $x_2 \in \mathbb{R}^{n_2}$, $f_1 : \mathbb{R}^{n_1} \times \mathbb{R}^{n_2}$ and $f_1(0, x_2) = 0$ for all $x_2 \in \mathbb{R}^{n_2}$. Moreover, we assume that $f_1(\cdot, x_2)$ is locally Lipschitz in x_1 for every x_2 and $f_2(x_1, \cdot)$ is locally Lipschitz in $x_2 \in \mathbb{R}^{n_2}$ for every $x_1 \in \mathcal{D}$. Then, the system (A.3) - (A.4) is exponentially stable with respect to $x_1 \in \mathbb{R}^{n_1}$ uniformly in x_{20}, see W. Haddad and V. Chellaboina (2011)[Chapter 4, Theorem 4.1 vii]. •

Note that Lemma 11 is adapted from W. Haddad and V. Chellaboina (2011)[Chapter 4, Theorem 4.1 vii)] and is a partial stability result.

A.2 Input-to-state stability

The stability analysis of systems with external input can be characterized by the so called input-to-state stability (ISS) property which was introduced by Eduardo Sontag in case of continuous-time systems. The following results are taken from D. Laila and A. Astolfi (2004); H. Edwards and Y. Lin and Y. Wang (2000); Z. Jiang and Y. Wang (2002).

Consider a discrete-time system of the form

$$x(k+1) = f(k, x(k), d(k)), \quad x(0) = x_0 \tag{A.5}$$

where $x(k) \in \mathbb{R}^n$ and $d(k) \in \mathbb{R}^{n_d}$ refers to the system state and disturbance at sampling instant $k \in \mathbb{N}$. We assume that i.) $0 = f(k, 0, 0)$ for all $k \in \mathbb{N}$ and ii.) the function f is continuous in its last two arguments for all $k \in \mathbb{N}$. The solution of (A.5) is denoted as $x(k, x_0, d_k)$ with input sequence $d_k = [d(0), d(1), \ldots, d(k)]$ at sampling instant $k \in \mathbb{N}$.

Definition 9. A function $\alpha : \mathbb{R}_+ \to \mathbb{R}_+$ is of class $\mathcal{K}$ if it is continuous, positive definite, and strictly increasing, and is of class $\mathcal{K}_\infty$ if it is also unbounded. A function $\beta : \mathbb{R}_+ \times \mathbb{N} \to \mathbb{R}_+$ is said to be of class $\mathcal{KL}$ if for each fixed $k \geq 0$, $\beta(\cdot, k)$ is of class $\mathcal{K}$, and for each fixed $s \geq 0$, $\beta(s, k)$ decreases to 0 as $k \to \infty$, see H. Edwards and Y. Lin and Y. Wang (2000). •

Definition 10. Uniform ISS. System (A.5) is called uniform ISS, if there exist functions $\gamma \in \mathcal{K}$ and $\beta \in \mathcal{KL}$ such that for any initial condition $x(0)$, each bounded input sequence d_k, it holds that

$$||x(k, x_0, d_k)|| \leq \beta(||x_0||, t) + \gamma(||d_k||_\infty) \tag{A.6}$$

for any $k \in \mathbb{N}$. •

Definition 11. Uniform ISS Lyapunov function. A continuous time-varying function $V : \mathbb{N} \times \mathbb{R}^n \to \mathbb{R}_+$ is called uniform ISS Lyapunov function, if there exists $\mathcal{K}_\infty$ functions $\alpha_1, \alpha_2, \alpha_3$ and a $\mathcal{K}$ function γ such that

$$\alpha_1(||x||) \leq V(k,x) \leq \alpha_2(||x||) \tag{A.7}$$
$$V(k+1, f(k,x,d)) - V(k,x(k)) \leq -\alpha_3(||x||) + \gamma(||d||) \tag{A.8}$$

hold for any $x \in \mathbb{R}^n$, $d \in \mathbb{R}^{n_d}$ and $k \in \mathbb{N}$. •

Lemma 12. The time-varying system (A.5) is uniform ISS if and only if it admits a uniform ISS-Lyapunov function. •

For the proof, we refer the reader to H. Edwards and Y. Lin and Y. Wang (2000).

B
Proofs

B.1 Proof of Lemma 1

Motivated by (C. Feller, 2017, C.6 Lemma 3.3), we apply Taylor's Theorem to eq. (2.14). Thus, we know that $\hat{B}_G(z,k) = \hat{B}_G(0,k) + [\nabla \hat{B}_G(0,k)]^\top z + \frac{1}{2} z^\top \nabla^2 \hat{B}_G(hz,k)z$ for some $h \in (0,1)$. Based on Definition 3, we know that $\hat{B}_G(0,k) = 0$ and $\nabla \hat{B}_G(0,k) = 0$. This results in

$$\hat{B}_G(z,k) = \frac{1}{2} z^\top \nabla^2 \hat{B}_G(hz,k)z \tag{B.1}$$

for some $h \in (0,1)$. Thus, there must exist a global upper bound $M \in \mathbb{S}^r_{++}$ on the Hessian, i.e., $\hat{B}_G(z,k) \leq \frac{1}{2} z^\top M z$ for all $z \in \mathbb{R}^r$ and $k \in \mathbb{N}$. Based on Definition 3 of the relaxed barrier functions, it holds that

$$\nabla^2 \hat{B}_G(z,k) = \begin{cases} \sum\limits_{i=1}^{q} C^{i\top} \frac{1}{(-C^i z + d^i(k))^2} C^i & -C^i z + d^i(k) > \delta \\ \sum\limits_{i=1}^{q} C^{i\top} \frac{1}{\delta^2} C^i & -C^i z + d^i(k) \leq \delta. \end{cases} \tag{B.2}$$

Moreover, it holds that $1/(-C^i z + d^i(k))^2 \leq 1/\delta^2$ for any $z \in \mathbb{R}^r$ and any $k, i \in \mathbb{N}$. Thus, we can conclude that the global upper bound on the Hessian is

$$\nabla^2 \hat{B}_G(z,k) \leq \frac{1}{\delta^2} C^\top C \tag{B.3}$$

for any $z \in \mathbb{R}^r$ and $k \in \mathbb{N}$. $\qquad\square$

B.2 Proof of Lemma 2

From Lemma 1, we know that

$$
\begin{aligned}
\hat{B}_{G,x}(x - x_{\text{ref}}(k), k) &\leq (x - x_{\text{ref}}(k))^\top M_x (x - x_{\text{ref}}(k)), \\
\hat{B}_{G,u}(u - u_{\text{ref}}(k), k) &\leq (u - u_{\text{ref}}(k))^\top M_u (u - u_{\text{ref}}(k)),
\end{aligned}
\tag{B.4}
$$

for all $x \in \mathbb{R}^n$, $u \in \mathbb{R}^m$ and $k \in \mathbb{N}$ with $M_x = \frac{1}{2\delta^2} C_x^\top C_x$ and $M_u = \frac{1}{2\delta^2} C_u^\top C_u$. If we now choose $u = u_{\text{ref}}(k) + K(x - x_{\text{ref}}(k))$, the second eq. of (B.4) transforms into $\hat{B}_{G,u}(K(x - x_{\text{ref}}(k)), k) \leq (x - x_{\text{ref}}(k))^\top K^\top M_u K(x - x_{\text{ref}}(k))$. Thus, we have that

$$
\hat{B}_K(x - x_{\text{ref}}(k), k) \leq (x - x_{\text{ref}}(k))^\top M (x - x_{\text{ref}}(k))
\tag{B.5}
$$

with $M = M_x + K^\top M_u K$, which proves the claimed statement. $\qquad\square$

B.3 Proof of Theorem 1

Owing to its positive definiteness, the basic idea is to choose $\hat{J}_N^*(x(k), k)$ as a Lyapunov function. The following proof consists of two parts.
A) First, we prove that there exist constants $a, b > 0$ such that the first eq. in (A.2) is satisfied. Observe that $\hat{J}_N^*(x(k), k)$ as defined in (2.15) depends on the sampling instance k, owing to the time-varying reference trajectory. Because $\hat{J}_N^*(x(k), k) > 0$, we can derive a lower bound with

$$
a||x(k) - x_{\text{ref}}(k))||^2 \leq \hat{J}_N^*(x(k), k),
\tag{B.6}
$$

with $a = \lambda_{\min}(Q) > 0$, where $\lambda_{\min}(Q)$ belongs to the minimal eigenvalue of the matrix $Q \in \mathbb{S}_{++}^n$. Using the inequality (2.35) of Lemma 2, we can derive a uniform global upper bound. We choose the terminal cost as

$$
F(x_N, k + N) = (x_N - x_{\text{ref}}(k + N))^\top P (x_N - x_{\text{ref}}(k + N)), \qquad P \in \mathbb{S}_{++}, \tag{B.7}
$$

where P is the solution of the modified discrete-time algebraic Riccati equation. In particular, we have

$$
P = A_K^\top P A_K + K^\top (R + \epsilon M_u) K + Q + \epsilon M_x
\tag{B.8a}
$$

$$
K = -(R + B^\top P B + \epsilon M_u)^{-1} B^\top P A,
\tag{B.8b}
$$

where $A_K = A + BK$, see also C. Feller (2017), and M_x, M_u is defined according to Lemma 2. More precisely, due to optimality, we have that

$$\hat{J}_N^*(x(k), k) \leq \hat{J}_N(x(k), k) = \sum_{j=0}^{N-1} ||x_j - x_{\text{ref}}(k+j)||_{Q+K^\top RK}^2$$
$$+ \epsilon \hat{B}_K(x_j - x_{\text{ref}}(k+j), k) \tag{B.9}$$
$$+ ||x_N - x_{\text{ref}}(k+N)||_P^2,$$

where $\hat{J}_N(x(k), k)$ belongs to the suboptimal value function with the suboptimal input sequence $u(x(k), k) = \{u_{\text{ref}}(k) + K(x_0 - x_{\text{ref}}(k)), \ldots, u_{\text{ref}}(k+N-1) + K(x_{N-1} - x_{\text{ref}}(k+N-1))\}$. Based on Assumption 3 and the suboptimal input sequence, we have that

$$x_j - x_{\text{ref}}(k+j) = A_K^j(x_0 - x_{\text{ref}}(k)). \tag{B.10}$$

Together with the upper bound from Lemma 2 with $\hat{B}_K(x_j - x_{\text{ref}}(k+j), k) \leq (x_j - x_{\text{ref}}(k+j))^\top M(x_j - x_{\text{ref}}(k+j))$ where $M \in \mathbb{S}_{++}^n$, eq. (B.9) transforms into

$$\hat{J}_N^*(x(k), k) \leq \sum_{j=0}^{N-1} ||A_K^j(x_0 - x_{\text{ref}}(k))||_{Q+K^\top RK}^2 + \epsilon ||A_K^j(x_0 - x_{\text{ref}}(k)))||_M^2$$
$$+ ||A_K^N(x_0 - x_{\text{ref}}(k))||_P^2$$
$$= (x_0 - x_{\text{ref}}(k))^\top \Omega(x_0 - x_{\text{ref}}(k))$$
$$\leq b||x_0 - x_{\text{ref}}(k)||_\Omega^2 \tag{B.11}$$

with $b = \lambda_{\max}(\Omega) > 0$ where $\lambda_{\max}(\Omega)$ is the maximal eigenvalue of the matrix $\Omega = \sum_{j=0}^{N-1}(A_K^j)^\top[Q + K^\top RK + \epsilon M]A_K^j + (A_K^N)^\top PA_K^N \in \mathbb{S}_{++}^n$.

B) Second, we show that the value function decreases along two consecutive sampling instants. We denote by $\hat{J}_N(x(k+1), k+1)$, the suboptimal cost at the successor state $x(k+1)$, and the successor reference trajectory $\{x_{\text{ref}}(k+1+j)\}_{j=0,\ldots,N}$, $\{u_{\text{ref}}(k+1+j))\}_{j=0,\ldots,N-1}$ at sampling instant $k+1$. Moreover, $\hat{J}_N^*(x(k), k)$ denotes the optimal cost at the system state $x(k)$ and reference trajectory $\{x_{\text{ref}}(k+j)\}_{j=0,\ldots,N}$, $\{u_{\text{ref}}(k+j))\}_{j=0,\ldots,N-1}$, both at sampling instant k. Notice that, for any $x(k) \in \mathbb{R}^n$ and sampling instant $k \in \mathbb{N}$, there exist optimal input and state sequences $u^*(x(k), k) = \{u_0^*, \ldots, u_{N-1}^*\}$ and $x^*(x(k), k) = \{x_0^*, \ldots, x_N^*\}$. The suboptimal input and state sequences are defined with (B.8b) as $u^+(x(k), k+1) = \{u_1^*, \ldots, u_{N-1}^*, u_N^+\}$ and $x^+(x(k), k+1) = \{x_1^*, \ldots, x_N^*, Ax_N^* + Bu_N^+\}$ with $u_N^+ = u_{\text{ref}}(k+N) + K(x_N^* - x_{\text{ref}}^*(k+N))$ and $x_N^+ = Ax_N^* + Bu_N^+$. By simply using the fact that $\hat{J}_N(x(k+1), k+1)$ is suboptimal, the inequality

$$\hat{J}_N^*(x(k+1), k+1) - \hat{J}_N^*(x(k), k) \leq \hat{J}_N(x(k+1), k+1) - \hat{J}_N^*(x(k), k) \quad \text{(B.12a)}$$

with $\hat{J}_N(x(k+1), k+1) - \hat{J}_N^*(x(k), k) = \sum_{j=1}^{N} \ell(x_j^+, u_j^+, k+j) + (x_N^* - x_{\text{ref}}(k+N))A_K^\top PA_K(x_N^* - x_{\text{ref}}(k+N)) - \sum_{j=0}^{N-1} \ell(x_j^*, u_j^*, k+j) + (x_N^* - x_{\text{ref}}(k+N))P(x_N^* - x_{\text{ref}}(k+N))$ holds. Now we can apply the results of Lemma 2 to uniformly globally upper bound $\ell(x_N^+, u_N^+; r(k+1)) = \|x_N^* - x_{\text{ref}}(k+N)\|_Q^2 + \|K(x_N^* - x_{\text{ref}}(k+N))\|_R^2 + \epsilon \hat{B}_K(x_N^* - x_{\text{ref}}(k+N), k)$, where $\hat{B}_K(x_N^* - x_{\text{ref}}(k+N), k) = \hat{B}_{G,x}(x_N^* - x_{\text{ref}}(k+N), k) + \hat{B}_{G,u}(K(x_N^* - x_{\text{ref}}(k+N)), k)$, which leads to

$$\ell(x_N^+, u_N^+, k+1) \leq \|(x_N^* - x_{\text{ref}}(k+N))\|_Q^2 + \|K(x_N^* - x_{\text{ref}}(k+N))\|_R^2 \\ + \epsilon(x_N^* - x_{\text{ref}}(k+N))\tilde{K}(x_N^* - x_{\text{ref}}(k+N)). \quad \text{(B.13)}$$

with $\tilde{K} = M_x + K^\top M_u K$. Thus, we result in

$$\hat{J}_N^*(x(k+1), k+1) - \hat{J}_N^*(x(k), k) \leq -\hat{\ell}(x(k), u(k), k) \\ + (x_N^* - x_{\text{ref}}(k+N))^\top P_0(x_N^* - x_{\text{ref}}(k+N)), \quad \text{(B.14)}$$

where $P_0 = (A_K^\top PA_K + K^\top(R + \epsilon M_u)K + Q + \epsilon M_x - P)$. By definition of the modified Riccati equation defined in (B.8a), P_0 is equal to 0. Using the inequality $\hat{\ell}(x(k), u(k), k) \geq \ell(x(k), u(k), k)$ with $\ell(x(k), u(k), k) = \|x(k) - x_{\text{ref}}(k)\|_Q^2 + \|u(k) - u_{\text{ref}}(k)\|_R^2$ and $\ell(x(k), u(k), k) \geq \|x(k) - x_{\text{ref}}(k)\|_Q^2$, we obtain

$$\hat{J}_N^*(x(k+1), k+1) - \hat{J}_N^*(x(k), k) \leq -\ell(x(k), u(k), k) \\ \leq -c\|x(k) - x_{\text{ref}}(k)\|^2 \quad \text{(B.15)}$$

with $c = \lambda_{\min}(Q) > 0$ where $\lambda_{\min}(Q)$ corresponds to the minimal generalized eigenvalue of the matrix $Q \in \mathbb{S}_{++}^n$. According to Lemma 10, we have exponential stability of the closed-loop system (2.18) for the origin which implies asymptotic stability. $\qquad\square$

B.4 Proof of Theorem 2

We now use the results from **A)** and **B)** from the proof of Theorem 1 to show that the state and input constraint set $x(k) \in \mathcal{X}$ and $u(k) \in \mathcal{U}$ will be satisfied for all $k \in \mathbb{N}$ if $x(0) \in \hat{E}_N(\delta)$. The following proof will consist of two parts. Firstly, we show that the input and state constraints will be satisfied initially if $x(0) \in \hat{E}_N(\delta)$. Secondly, we will show that if $x(0) \in \hat{E}_N(\delta)$, then the system state and control input constraints will be satisfied for all future sampling instants $k \in \mathbb{N}$. The proof is similar to Theorem 3.6 and Corollary 3.2 in C. Feller (2017).

Part 1: As shown in Section 2.2, the reference trajectory is globally uniformly exponentially stable for the closed-loop system. Thus, the value function $\hat{J}_N^*(x(k), k)$ is strictly monotonically decreasing for all $k \in \mathbb{N}$. Thus, we have that $\lim_{k \to \infty} \hat{J}_N^*(x(k), k) = 0$. By summing up all future sampling instants of (B.14), we obtain

$$\sum_{i=k}^{\infty} \hat{J}_N^*(x(i), i) - \hat{J}_N^*(x(i+1), i+1) \geq \sum_{i=k}^{\infty} \ell(x(i), u(i), i) \qquad (B.16)$$

where $x(k)$ and $u(k)$ belong to the actual realizations of the state and input of the closed-loop system (3.4) at sampling instant k. Owing to the left-hand side being a telescopic sum, eq. (B.16) transforms into $\hat{J}_N^*(x(k), k) \geq \sum_{i=k}^{\infty} \ell(x(i), u(i), i)$. By the property of the unconstrained optimal control problem, we have, due to optimality, $\sum_{i=k}^{\infty} \ell(x(i), u(i), i) = ||x(i) - x_{\text{ref}}(k)||_Q^2 + ||u(i) - u_{\text{ref}}(k)||_R^2 \geq J_\infty^*(x(k), k)$. Thus, with the definition of $\sum_{i=k}^{\infty} \hat{\ell}(x(i), u(i), i) = \sum_{i=k}^{\infty} ||x(i) - x_{\text{ref}}(i)||_Q^2 + ||u(i) - u_{\text{ref}}(i)||_R^2 + \epsilon \sum_{i=k}^{\infty} \hat{B}_{G,\text{xu}}(x(i), u(i), i)$, for any sampling instant $k \in \mathbb{N}$, the following inequality

$$\hat{J}_N^*(x(k), k) \geq J_\infty^*(x(k), k) + \epsilon \sum_{i=k}^{\infty} \hat{B}_G(x(i), u(i), i) \qquad (B.17)$$

holds. Based on eq. (B.17) and the relaxed barrier function $\hat{B}_{G,\text{xu}}(\cdot, \cdot, \cdot)$ being strictly positive, we obtain for any $k \in \mathbb{N}$ and $\epsilon \in \mathbb{R}_{++}$ that

$$\epsilon \hat{B}_G(x(k), u(k), k) \leq \hat{J}_N^*(x(k), k) - J_\infty^*(x(k), k). \qquad (B.18a)$$

If we now define the set $\hat{\mathcal{E}}_N(\delta)$ as

$$\hat{\mathcal{E}}_N(\delta) = \{x \in \mathbb{R}^n | \hat{J}_N^*(x, 0) - J_\infty^*(x, 0) \leq \bar{\beta}(\delta)\} \qquad (B.19)$$

and let $x(0) \in \hat{\mathcal{E}}_N(\delta)$, then $\epsilon \hat{B}_G(x(k), u(k), k) \leq \bar{\beta}(\delta)$, which, by the definition of $\bar{\beta}(\delta)$, implies that we have exact constraint satisfaction in state and input for $k = 0$.

Part 2: Next, we prove that $\alpha(k) = \hat{J}_N^*(x(k),k) - J_\infty^*(x(k),k)$ decreases monotonically in time for any $k \in \mathbb{N}$. We begin with the consideration of the difference of $\alpha(k)$ at two consecutive sampling instants, which leads together with $\hat{J}_N^*(x(k+1),k+1) - \hat{J}_N^*(x(k),k) \leq \hat{\ell}(x(k),u(k),k)$ to

$$\alpha(k+1) - \alpha(k) \leq -\hat{\ell}(x(k),u(k),k) + J_\infty^*(x(k),k) - J_\infty^*(x(k+1),k+1). \quad \text{(B.20)}$$

Now recall $\hat{\ell}(x(k),u(k),k) = ||x(k) - x_{\text{ref}}(k)||_Q^2 + ||u(k) - u_{\text{ref}}(k)||_R^2$ $+ \hat{B}_G(x(k),u(k),k)$. Due to the principle of optimality, we have $J_\infty^*(x(k),k) - J_\infty^*(x(k+1),k+1) \leq ||x(k) - x_{\text{ref}}(k)||_Q^2 + ||u(k) - u_{\text{ref}}(k)||_R^2$. Thus, we obtain

$$\alpha(k+1) - \alpha(k) \leq -\epsilon\left(\hat{B}_{G,\text{xu}}(x(k),u(k),k)\right) \quad \text{(B.21)}$$

Given that the terms on the right-hand side are negative definite, we can conclude that $\alpha(k) \leq \alpha(0)$ for any $k \in \mathbb{N}$. Thus, if $x(0) \in \hat{\mathcal{E}}_N(\delta)$, it holds that $x(k) \in \hat{\mathcal{E}}_N(\delta)$ for all $k \in \mathbb{N}$. Moreover, we know that based on eq. (B.17) and $\alpha(k) = 0$ for $x(k) = x_{\text{ref}}(k)$ and $u(k) = u_{\text{ref}}(k)$, that $\alpha(k)$ is positive definite for any $k \in \mathbb{N}$. Thus, we can conclude that, owing to ϵ and $\bar{\beta}(\delta)$ being strictly positive, the set $\hat{\mathcal{E}}_N(\delta)$ is non-empty. Furthermore, due to $\alpha(k)$ converging to 0 (see eq. (B.21)), we can conclude that the set $\hat{\mathcal{E}}_N(\delta)$ is compact. $\qquad\square$

B.5 Proof of Lemma 4

We follow the ideas of (C. Feller, 2017, C.6 Lemma 3.3). Applying Taylor's Theorem to the first argument of eq. (2.28), we know that $\hat{B}_G(z,\theta) = \hat{B}_G(0,\theta) + [\nabla\hat{B}_G(0,\theta)]^\top z + \frac{1}{2}z^\top\nabla^2\hat{B}_G(hz,\theta)z$ for some $h \in (0,1)$. Based on Definition 4, we know that $\hat{B}_G(0,\theta) = 0$ and $\nabla\hat{B}_G(0,\theta) = 0$. Hence, we obtain

$$\hat{B}_G(z,\theta) = \frac{1}{2}z^\top\nabla^2\hat{B}_G(hz,\theta)z \quad \text{(B.22)}$$

for some $h \in (0,1)$. Based on Definitions 1, 2 and 4 of the relaxed barrier functions, it holds that

$$\nabla^2\hat{B}_G(z,\theta) = \begin{cases} \sum\limits_{i=1}^{q} C^{i\top}\frac{1}{(-C^i z + d^i(\theta))^2}C^i & G(\theta) > \delta \\ \sum\limits_{i=1}^{q} C^{i\top}\frac{1}{\delta^2}C^i & G(\theta) \leq \delta \end{cases} \quad \text{(B.23)}$$

with $G(\theta) = -C^i z + d^i(\theta)$. Obviously, it holds that $1/(-C^i z + d^i(\theta))^2 \leq 1/\delta^2$ for any $z \in \mathbb{R}^r$ and any $i \in \mathbb{N}, \theta \in \mathbb{R}$. Thus, we can conclude that

$$\nabla^2 \hat{B}_G(z, \theta) \leq \frac{1}{\delta^2} C^\top C \qquad (B.24)$$

for any $(z, \theta) \in \mathbb{R}^r \times \mathbb{R}$, which proves the claimed statement. $\qquad \square$

B.6 Proof of Lemma 5

From Lemma 4, we know that

$$\hat{B}_{G,x}(x - x_{\text{ref}}(\theta), \theta) \leq (x - x_{\text{ref}}(\theta))^\top M_x (x - x_{\text{ref}}(\theta)) \qquad (B.25)$$

$$\hat{B}_{G,u}(u - u_{\text{ref}}(\theta), \theta) \leq (u - u_{\text{ref}}(\theta))^\top M_u (u - u_{\text{ref}}(\theta)) \qquad (B.26)$$

for all $(x, u, \theta) \in \mathbb{R}^n \times \mathbb{R}^m \times \mathbb{R}$. If we now choose $u = u_{\text{ref}}(\theta) + K(x - x_{\text{ref}}(\theta))$, (B.26) transforms into $\hat{B}_{G,u}(u - u_{\text{ref}}(\theta), \theta) \leq (x - x_{\text{ref}}(\theta))^\top K^\top M_u K(x - x_{\text{ref}}(\theta))$, which together with (B.25) proves the claimed statement. $\qquad \square$

B.7 Proof of Theorem 3

The basic idea of the proof is that $\hat{J}_N^*(x(k), \theta(k))$ is a Lyapunov candidate that fulfills the conditions in (A.2). The following proof of the theorem consists of two parts.

Part A) We first prove that $\hat{J}_N^*(x(k), \theta(k))$ has uniform lower and upper bounds $a||x(k) - x_{\text{ref}}(\theta(k))||^\sigma$ and $b||x(k) - x_{\text{ref}}(\theta(k))||^\sigma$ for $\sigma = 2$ with respect to the first variable. Owing to the overall cost $\hat{J}_N^*(x(k), \theta(k)) > 0$, we can conclude that

$$\hat{J}_N^*(x(k), \theta(k)) \geq a||x(k) - x_{\text{ref}}(\theta(k))||^2, \qquad (B.27)$$

with $a = \lambda_{\min}(Q) > 0$ where $\lambda_{\min}(Q)$ is the minimal eigenvalue of the matrix $Q \in \mathbb{S}_{++}^n$. Moreover, we can also derive an uniform upper bound for the chosen Lyapunov candidate $\hat{J}_N^*(x(k), \theta(k))$ for all $x(k) \in \mathbb{R}^n, \theta \in \mathbb{R}$. By applying the suboptimal input sequence $u(x(k), \theta(k)) = \{u_{\text{ref}}(k) + K(x_0 - x_{\text{ref}}(k)), \ldots, u_{\text{ref}}(k + N - 1) + K(x_{N-1} - x_{\text{ref}}(k + N - 1))\}$ with $x_j = x_{\text{ref}}(k + j) + (A + BK)^j(x_0 - x_{\text{ref}}(k))$, we get the suboptimal cost $\hat{J}_N(x(k), \theta(k))$. By optimality it holds that $\hat{J}_N^*(x(k), \theta(k)) \leq \hat{J}_N(x(k), \theta(k))$. Using the quadratic upper bounds from Lemma 5, we know that $\hat{J}_N(x(k), \theta(k)) \leq (x(k) - x_{\text{ref}}(k))^\top \Omega(x(k) - x_{\text{ref}}(k)) \leq b||x(k) -$

$x_{\text{ref}}(k)||^2$ where $b = \lambda_{\max}(\Omega) > 0$ refers to the maximal eigenvalue of $\Omega = \sum_{j=0}^{N-1} A_K^{j\,\top}(Q + \varepsilon M + K^\top R K)A_K^j + A_K^{N\,\top} P A_K^N$ and $A_K = A + BK$.

Part **B)** Second, we show that value function $\hat{J}_N^*(x(k), \theta(k))$ decreases along the reference path defined in Assumption 4 with respect to the first variable. Considering now the difference of the Lyapunov function candidate $\hat{J}_N^*(x(k), \theta(k))$ at two consecutive sampling instants, we have

$$\hat{J}_N^*(x(k+1), \theta(k+1)) - \hat{J}_N^*(x(k), \theta(k)). \tag{B.28}$$

The corresponding open-loop optimal control sequences for any $x(k) \in \mathbb{R}^n$, $\theta(k) \in \mathbb{R}_+$ and $k \in \mathbb{N}$ are defined as $v^*(x(k), \theta(k)) = \{v_0^*, \dots, v_{N-1}^*\}$, $u^*(x(k), \theta(k)) = \{u_0^*, \dots, u_{N-1}^*\}$ and $x^*(x(k), \theta(k)) = \{x_0^*, \dots, x_N^*\}$. Let us define the suboptimal cost as $\hat{J}_N(x(k+1), \theta(k+1))$ with the corresponding suboptimal sequences defined as

$$\begin{aligned}
v^+(x(k), \theta(k)) &= \{v_1^*, \dots, v_{N-1}^*, 0\} \\
u^+(x(k), \theta(k)) &= \{u_1^*, \dots, u_{N-1}^*, u_N^+\} \\
x^+(x(k), \theta(k)) &= \{x_1^*, \dots, x_N^*, Ax_N^* + Bu_N^+\} \\
\theta^+(x(k), \theta(k)) &= \{\theta_1^*, \dots, \theta_N^*, \theta_N^* + 1\}
\end{aligned}$$

with $u_N^+ = u_{\text{ref}}(\theta_N^*) + K(x_N^* - x_{\text{ref}}(\theta_N^*))$. Note that based on Assumption 2, the suboptimal but stabilizing solution always exists. Due to optimality, we can reformulate (B.28) into $\hat{J}_N^*(x(k+1), \theta(k+1)) - \hat{J}_N^*(x(k), \theta(k)) \leq \hat{J}_N(x(k+1), \theta(k+1)) - \hat{J}_N^*(x(k), \theta(k))$. Moreover, considering the definition of the suboptimal cost $\hat{J}_N(x(k+1), \theta(k+1))$, we further get

$$\begin{aligned}
&\hat{J}_N(x(k+1), \theta(k+1)) - \hat{J}_N^*(x(k), \theta(k)) \\
&= \sum_{j=1}^{N} \hat{\ell}(x_j^+, u_j^+, \theta_j^+) + \hat{F}(x_N^+, \theta_N^+) \\
&\quad - \sum_{j=0}^{N-1} \hat{\ell}(x_j^*, u_j^*, \theta_j^*) - \hat{F}(x_N^*, \theta_N^*)
\end{aligned}$$

with $\hat{\ell}(x_j^+, u_j^+, \theta_j^+) = ||x_j^+ - x_{\text{ref}}(\theta_j^+)||_Q^2 + ||u_j^+ - u_{\text{ref}}(\theta_j^+)||_R^2 + \varepsilon \hat{B}_{\text{G,xv}}(x_j^+, u_j^+, \theta_j^+, v_j^+)$, $\hat{F}(x_N^+, \theta_N^+) = ||x_N^+ - x_{\text{ref}}(\theta_N^+)||_P^2$, $\hat{\ell}(x_j^*, u_j^*, \theta_j^*) = ||x_j^* - x_{\text{ref}}(\theta_j^*)||_Q^2 + ||u_j^* - u_{\text{ref}}(\theta_j^*)||_R^2 +$

$\varepsilon \hat{B}_{G,xv}(x_j^*, u_j^*, \theta_j^*, v_j^*)$ and $F(x_N^*, \theta_N^*) = ||x_N^* - x_{\text{ref}}(\theta_N^*)||_P^2$. Thus, we have that

$$
\begin{aligned}
&\hat{J}_N(x(k+1), \theta(k+1)) - \hat{J}_N(x(k), \theta(k)) \\
&= -\hat{\ell}(x(k), u(k), \theta(k)) + \hat{\ell}(x_N^+, u_N^+, \theta_N^+) \\
&\quad + \hat{F}(x_{N+1}^+, \theta_{N+1}^+) - \hat{F}(x_N^*, \theta_N^*) \\
&= -\hat{\ell}(x(k), u(k), \theta(k)) + ||A_K(x_N^* - x_{\text{ref}}(\theta_N^*))||_Q^2 \\
&\quad + ||K(x_N^* - x_{\text{ref}}(\theta_N^*))||_R^2 + \varepsilon \hat{B}_K(x_N^+, \theta_N^+) + \varepsilon \hat{B}_G(v_N^+) \\
&\quad + ||A_K(x_N^* - x_{\text{ref}}(\theta_N^*))||_P^2 - ||x_N^* - x_{\text{ref}}(\theta_N^*)||_P^2 \\
&\leq -\hat{\ell}(x(k), u(k), \theta(k)) + \varepsilon \hat{B}_G(v_N^+ = 0) \\
&\quad + (x_N^* - x_{\text{ref}}(\theta_N^*))^\top (Q + \epsilon M_x + K^\top (R + \epsilon M_u)K + A_K^\top P A_K - P) \\
&\quad (x_N^* - x_{\text{ref}}(\theta_N^*)) \\
&= -\hat{\ell}(x(k), u(k), \theta(k)).
\end{aligned}
$$

Here the first inequality holds when we apply Lemma 5. The last equality holds from the choice of P to be the solution of the modified Riccati. We then use the property of the gradient recentered and relaxed barrier function, i.e., $\hat{\ell}(x(k), u(k), \theta(k)) \geq \ell(x(k), u(k), \theta(k)) = ||x(k) - x_{\text{ref}}(\theta(k))||_Q^2$ $+ ||u(k) - u_{\text{ref}}(\theta(k))||_R^2$ to get

$$
\begin{aligned}
&\hat{J}_N^*(x(k+1), \theta(k+1)) - \hat{J}_N^*(x(k), \theta(k)) \\
&\leq \ell(x(k), u(k), \theta(k)) \\
&\leq -c||x(k) - x_{\text{ref}}(\theta(k))||^2
\end{aligned}
\tag{B.30}
$$

with respect to the first variable with $c = \lambda_{\min}(Q) > 0$ where $\lambda_{\min}(Q)$ corresponds to the minimal eigenvalue of the matrix $Q \in \mathbb{S}_{++}^n$. In our case, the variables of Lemma 11 correspond to $(x_1(k), x_2(k)) = (x(k) - x_{\text{ref}}(\theta(k)), \theta(k))$. From **A)** and **B)**, we can conclude that the closed-loop system (2.32) - (2.33) is exponentially stable for the reference path $P(\theta)$ uniformly in $\theta \in \mathbb{R}$ which proves the claimed statement. $\qquad \square$

B.8 Proof of Theorem 4

We now use the results of Part **A)** and **B)** from the proof of Theorem 3 to show that the input, virtual input, and state constraints will be satisfied for all sampling instants k if $x(0) \in \hat{E}_N(\delta)$. First, we prove $\varepsilon \hat{B}_{G,xv}(x(k), u(k), \theta(k), v(k)) \leq \hat{J}_N^*(x(k), \theta(k)) - J_\infty^*(x(k), \theta(k))$ for all $k \in \mathbb{N}$. As shown in **B)**, the value function strictly monotonically decreases towards the reference path. Thus, we have that $\lim_{k\to\infty} \hat{J}_N^*(x(k), \theta(k)) = 0$. By summing up all future time steps of (B.30), we get

$$\sum_{i=k}^{\infty} \hat{J}_N^*(x(i+1), \theta(i+1)) - \hat{J}_N^*(x(i), \theta(i)) \geq \sum_{i=k}^{\infty} \ell(x(i), u(i), \theta(i)) \quad \text{(B.31)}$$

Due to the left-hand side being a telescopic sum, (B.31) turns into $\hat{J}_N^*(x(k), \theta(k)) \geq$

$\sum_{i=k}^{\infty} \ell(x(i), u(i), \theta(i))$. By optimality, we have $\sum_{i=k}^{\infty} \ell(x(i), u(i), \theta(i))$
$= \sum_{i=k}^{\infty} ||x(i) - x_{\text{ref}}(\theta(i))||_Q^2 + ||u(i) - u_{\text{ref}}(\theta(i))||_R^2 \geq J_\infty^*(x(k), \theta(k))$. Thus, with the definition of $\sum_{i=k}^{\infty} \ell(x(i), u(i), \theta(i)) = \sum_{i=k}^{\infty} ||x(i) - x_{\text{ref}}(\theta(i))||_Q^2 + ||u(i) - u_{\text{ref}}(i)||_R^2 +$
$\epsilon \sum_{i=k}^{\infty} \hat{B}_{G,xv}(x(i), u(i), \theta(i), v(i))$, it holds that for each time step, we have the following inequality

$$\begin{aligned} \hat{J}_N^*(x(k), \theta(k)) \geq & J_\infty^*(x(k), \theta(k)) \\ & + \epsilon \sum_{i=k}^{\infty} \hat{B}_{G,xv}(x(i), u(i), \theta(i), v(i)) \end{aligned} \quad \text{(B.32)}$$

with $\hat{B}_{G,xv}(x(i), u(i), \theta(i), v(i)) = \hat{B}_{G,x}(x(i) - x_{\text{ref}}(\theta(i)), \theta(i))$
$+ \hat{B}_{G,u}(u(i) - u_{\text{ref}}(\theta(i)), \theta(i)) + \hat{B}_{G,v}(v(i))$. Owing to eq. (B.32) and the relaxed barrier function $\hat{B}_G(\cdot, \cdot)$, $\hat{B}_G(\cdot)$ being non-negative, we obtain for any $k \in \mathbb{N}$

$$\epsilon \hat{B}_{G,xv}(x(k), u(k), \theta(k), v(k)) \leq \alpha(k). \quad \text{(B.33)}$$

with $\alpha(k) = \hat{J}_N^*(x(k), \theta(k)) - J_\infty^*(x(k), \theta(k))$. If we define the set $\hat{\mathcal{E}}_N(\delta, \theta(0))$ as

$$\hat{\mathcal{E}}_N(\delta, \theta(0)) = \{x \in \mathbb{R}^n | \hat{J}_N^*(x, \theta(0)) - J_\infty^*(x, \theta(0)) \leq \epsilon \bar{\beta}(\delta)\} \quad \text{(B.34)}$$

and let $x(0) \in \hat{\mathcal{E}}_N(\delta, \theta(0))$ (with $x(0)$ being the initial state of the system state trajectory), then $\hat{B}_{G,xv}(x(k), u(k), \theta(k), v(k)) \leq \bar{\beta}(\delta)$, which by definition of $\bar{\beta}(\delta)$ implies that we have exact constraint satisfaction in state, input, and virtual control input for $k = 0$.

Next, we prove that $\alpha(k) = \hat{J}_N^*(x(k), \theta(k)) - J_\infty^*(x(k), \theta(k))$ decreases monotonically in time. Thus, we consider the difference between two consecutive sampling instants, together with (B.30), we have that $\alpha(k+1) - \alpha(k) \leq$
$- \ell(x(k), u(k), \theta(k)) + J_\infty^*(x(k), \theta(k)) - J_\infty^*(x(k+1), \theta(k+1))$. Now recall

$\ell(x(k), u(k), \theta(k)) = ||x(k) - x_{\text{ref}}(\theta(k))||_Q^2 + ||u(k) - u_{\text{ref}}(\theta(k))||_R^2 + \epsilon \hat{B}_{G,x}(x(k) - x_{\text{ref}}(\theta(k)), \theta(k)) + \epsilon B_{G,u}(u(k) - u_{\text{ref}}(\theta(k)), \theta(k)) + \epsilon \hat{B}_{G,v}(v(k))$. Due to the principle of optimality, we have $J_\infty^*(x(k), \theta(k)) - J_\infty^*(x(k+1), \theta(k+1)) \leq ||x(k) - x_{\text{ref}}(\theta(k))||_Q^2 + ||u(k) - u_{\text{ref}}(\theta(k))||_R^2$. Thus, we obtain

$$\alpha(k+1) - \alpha(k) \leq -\epsilon \Big(\hat{B}_{G,x}(x(k) - x_{\text{ref}}(\theta(k)), \theta(k))$$
$$+ \hat{B}_{G,u}(u(k) - u_{\text{ref}}(\theta(k)), \theta(k)) + \hat{B}_{G,v}(v(k)) \Big) \tag{B.35}$$

Since the terms on the right-hand side are negative definite, we can conclude that $\alpha(k) \leq \alpha(0)$ for any $k \in \mathbb{N}$. Thus if $x(0) \in \hat{\mathcal{E}}_N(\delta, \theta(0))$ for some $\theta(0) \in \mathbb{R}$, it holds that $x(k) \in \hat{\mathcal{E}}_N(\delta, \theta(0))$ for all $k \in \mathbb{N}$ and thus it holds that the constraints are satisfied. Moreover, we know that based on eq. (B.32) and the definition of the gradient recentered barrier function (see Definition 4), we know that $\alpha(k)$ is non-negative for all $k \in \mathbb{N}$. Thus, we can conclude that, owing to ϵ and $\bar{\beta}(\delta)$ being strictly positive, that the set $\hat{\mathcal{E}}_N(\delta, \theta(0))$ is non-empty. Furthermore, owing to $\alpha(k)$ being convergent (see eq. (B.35)), we can conclude that the set $\hat{\mathcal{E}}_N(\delta, \theta(0))$ is compact. □

B.9 Proof of Theorem 5

These results are rather standard and follow from LQR theory or multiparametric programming or directly by calculations as outlined below.

In the following, we consider k arbitrary but fixed. Notice that (3.3) can be written as minimize $w^\top H(x)w$ subject to $C(k)w = B(k)x$ and $w^\top = [u_0^\top \ x_1^\top \ \ldots \ u_{N-1}^\top \ x_N^\top]$. Further, it can be verified that $C(k)$ has full row rank. Hence, after relabeling variables, we can partition the equality constraint into $C(k)w = C_1(k)w_1 + C_2(k)w_2 = B(k)x$ with $\det(C_1(k)) \neq 0$ for any $k \in \mathbb{N}$, $w_1^\top = [x_1^\top \ \ldots \ x_N^\top]$ and $w_2^\top = [u_0^\top \ \ldots \ u_{N-1}^\top]$. Hence the variable w_1 can be eliminated and the problem can be transformed into an unconstrained quadratic and convex optimization problem in w_2. More precisely, the equality constrained optimization problem is denoted as $w_1^\top H_1(x, \epsilon)w_1 + w_2^\top H_2 w_2$ subject to $C_1(k)w_1 + C_2(k)w_2 = B(k)p$. The equality constraint can be rewritten into $w_1 = C_1(k)^{-1}(B(k)x - C_2(k)w_2)$ and inserted into the objective function. Hence, we result into an unconstrained quadratic and convex optimization problem which is associated with the following objective function $w_2^\top (H_2 + \tilde{H}_2(x, p_1))w_2 - 2w_2^\top \tilde{F}_2(x, p_1)B(k)x + c_2(x, p_1)$, where $\tilde{H}_2(x, p_1) = C_2(k)^\top C_1(k)^{-\top} H_1(x, \epsilon)C_1(k)^{-1}C_2(k)$, $\tilde{F}_2(x, p_1) = C_2(k)^\top C_1(k)^{-\top} H_1(x, \epsilon)C_1(k)^{-1}$ and

$c_2(x, p_1) = x^\top B(k)^\top C_1(k) H_1(x, \epsilon) C_1(k)^{-1} B(k) x$. From first order optimality conditions, it holds that

$$w_2^*(x, p_1) = (H_2 + \tilde{H}_2(x, p_1))^{-1} \tilde{F}_2(x, p_1) B(k) x \tag{B.36}$$

with $H_2 > 0$, $\tilde{H}_2(x, p_1) \geq 0$ for all $x \in \mathbb{R}^n$, $k \in \mathbb{N}$, $\epsilon > 0$ with $p_1 = [k, \epsilon]^\top$, where $w_2^*(x, p_1)$ corresponds to the optimal solution of the objective function associated with the unconstrained quadratic and constrained optimization problem for some x, p_1. From eq. (B.36), it follows that the solution of $V_1(x, p_1)$ is parameterized in x in the sense of $x_i(x, p_1) = K_{1,i}(x, p_1) x$ for $i = 1, \ldots, N$ and $u_i(x, p_1) = K_{3,i}(x, p_1) x$ for $i = 0, \ldots, N - 1$. $\qquad\square$

B.10 Proof of Lemma 7

Proof by contradiction. Suppose for all $\rho > 0$ there exists an $x \in \mathbb{R}^N$, $k \in \mathbb{N}$, and $\epsilon > 0$ such that $x_N(x, p_1)^\top Q_N x_N(x, p_1) > V_2^*(x, p_2) + \epsilon\rho$. Then by (3.3), Assumption 15 a.) and $\sum_{i=0}^{N-1} x_i^*(x, p_1)^\top Q x_i^*(x, p_1) + u_i^*(x, p_1)^\top R u_i^*(x, p_1) \geq 0$ (due to $(R, Q > 0)$), we have $V_1^*(x, p_1) > \frac{\Gamma(x) V_2^*(x, p_2)}{\epsilon} + \rho \Gamma(x)$. On the other hand, we have
$V_1(x, p_1) = \sum_{i=0}^{N-1} \xi_i(x, p_2)^\top Q \xi_i(x, p_2) + v_i(x, p_2)^\top R v_i(x, p_2) + \frac{\Gamma(x) V_2^*(x, p_2)}{\epsilon}$, where $\{v_i(x, p_2)\}_{i=0}^{N-1}$, $\{\xi_i(x, p_2)\}_{i=0}^{N}$ is a solution of (3.10) and it is feasible for (3.3). By optimality, we must have $V_1^*(x, p_1) \leq V_1(x, p_1)$, hence

$$\begin{aligned}
\frac{\Gamma(x) V_2^*(x, p_2)}{\epsilon} + \rho \Gamma(x) &< \frac{\Gamma(x) V_2^*(x, p_2)}{\epsilon} \\
&+ \sum_{i=0}^{N-1} \xi_i(x, p_2)^\top Q \xi_i(x, p_2) + v_i(x, p_2)^\top R v_i(x, p_2)
\end{aligned} \tag{B.37}$$

must hold. However, (B.37) cannot be true for all $\rho > 0$ since $\Gamma(x) \geq c(\sum_{i=0}^{N-1} \|\xi_i(x, p_2)\|^2 + \|v_i(x, p_2)\|^2)$ and therefore there exists a $\rho > 0$ (independent of x, p_2) such that $\rho \Gamma(x) \geq \rho c(\sum_{i=0}^{N-1} \|\xi_i(x, p_2)\|^2 + \|v_i(x, p_2)\|^2) \geq \sum_{i=0}^{N-1} \xi_i(x, p_2)^\top Q \xi_i(x, p_2) + v_i(x, p_2)^\top R v_i(x, p_2)$. $\qquad\square$

B.11 Proof of Theorem 6

In the following, we choose the optimal value function $V_1^*(x(k), p_1(k))$ as a Lyapunov function. Let $\mu(x(k), p_1(k)) = \{\mu_0, \ldots, \mu_{N-1}\}$ be a suboptimal solution for (3.3) (according to Assumption 15 c.)) and let $x(x(k), p_1(k)) = \{x_0, \ldots, x_N\}$ be the corresponding predicted state sequence together with the suboptimal value function $V_1(x(k), p_1(k))$. The proof consists of two steps. In step a.) we will show that the Lyapunov function is uniformly upper and lower bounded according to eq. (A.7). In step b.) we will consider the difference of the Lyapunov function between two consecutive sampling instants such that eq. (A.8) is fulfilled.

a.) Due to the positive definiteness of the optimal value function, it holds that

$$b_1 ||x(k)||^2 \leq V_1^*(x(k), p_1(k)) \tag{B.38}$$

for all $k \in \mathbb{N}$ with $b_1 = \lambda_{\min}(Q) > 0$ where $Q \in \mathbb{S}_{++}^n$. Moreover, due to optimality, we have

$$
\begin{aligned}
V_1^*(x(k), p_1(k)) &\leq V_1(x(k), p_1(k)) \\
&= \sum_{i=0}^{N-1} x_i^\top Q x_i + \mu_i^\top R \mu_i + \frac{\Gamma(x(k))}{\epsilon(k)} x_N^\top Q_N x_N \\
&= x(k)^\top Q x(k) + \sum_{i=1}^{N-1} x_i^\top Q x_i \\
&\quad + \sum_{i=0}^{N-1} \mu_i^\top R \mu_i + \frac{\Gamma(x(k))}{\epsilon(k)} x_N^\top Q_N x_N.
\end{aligned} \tag{B.39}
$$

Due to Assumption 15 c.), we know that $\sum_{l=0}^{N-1} ||\mu_l||^2 \leq \gamma(||x(k)||)$. Hence, we can conclude that

$$\sum_{i=0}^{N-1} \mu_i^\top R \mu_i \leq b_0 \sum_{i=0}^{N-1} ||\mu_i||^2 \leq \gamma(||x(k)||) \tag{B.40}$$

with $b_0 = \lambda_{\max}(R) > 0$, $R \in \mathbb{S}_{++}^m$ and $\lambda_{\max}(R)$ is the maximal eigenvalue of the matrix R. In the following, we will refer in the latter to the eigenvalue inequality. Moreover, since Assumption 15 a.) holds, we have that $x_i = A_i(k)x(k) + \sum_{l=0}^{i-1} B_{i-1-l}(k)\mu_l$ for $i = 0, \ldots, N-1$. Applying Young's inequality together with the eigenvalue inequality yields to

$$||x_i||_Q^2 \leq b_1 ||x(k)||^2$$
$$+ b_2 \left(\sum_{l=1}^{i} A_{i-l}(k) B_{l-1}(k)) \mu_{l-1} \right)^{\top} \left(\sum_{l=1}^{i} A_{i-l}(k) B_{l-1}(k)) \mu_{l-1} \right) \quad \text{(B.41)}$$

with $b_1, b_2 > 0$. Furthermore, using the fact that the matrix sequences $\{A_i(k)\}_{k \in \mathbb{N}}$ for $i = 0, \ldots, N$ and $\{B_i(k)\}_{k \in \mathbb{N}}$ for $i = 0, \ldots, N-1$ are uniformly bounded together with Young's inequality, and applying again Assumption 15 c.), transforms eq. (B.41) into

$$||x_i||_Q^2 \leq b_1 ||x(k)||^2 + b_3 \gamma(||x(k)||) \quad \text{(B.42)}$$

with $b_1 > 0$ and $b_3 > 0$. From eq. (B.39), (B.40) and (B.42), we have that

$$V_1^*(x(k), p_1(k)) \leq b_1 ||x(k)||^2 + b_4 \gamma(||x(k)||) + \frac{\Gamma(x(k))}{\epsilon(k)} x_N^{\top} Q_N x_N \quad \text{(B.43)}$$

with $b_4 = b_3 + 1 > 0$. Due to Assumption 15 c.), we know that $||x_N||^2 \leq c\beta(k)$ with $c > 0$. Thus, together with the maximal eigenvalue inequality, we can upper bound eq. (B.43) by

$$V_1^*(x(k), p_1(k)) \leq b_1 ||x(k)||^2 + b_4 \gamma(||x(k)||) + b_5 \frac{\Gamma(x(k))}{\epsilon(k)} \beta(k) \quad \text{(B.44)}$$

with $b_5 > 0$. Together with $\Gamma(x(k)) = \alpha x(k)^{\top} x(k)$, $\alpha > 0$ and the choice of $\epsilon(k) > \beta(k)$ with $\lim_{k \to \infty} \epsilon(k) = 0$, we can conclude that

$$V_1^*(x(k), p_1(k)) \leq b_2 ||x(k)||^2 + b_4 \gamma(||x(k)||) \quad \text{(B.45)}$$

for any $k \in \mathbb{N}$ and some $b_2, b_4 > 0$. Hence, the condition of a time-invariant lower and upper bound is fulfilled.

b.) Now let the optimal value $V_1^* : \mathbb{R}^n \times \mathbb{N} \times \mathbb{R}_+ \to \mathbb{R}_+$ at sampling instant k be denoted as

$$V_1^*(x(k), p_1(k)) = \sum_{i=0}^{N-1} (x_i^*)^{\top} Q x_i^* + (u_i^*)^{\top} R u_i^* + \frac{\Gamma(x(k))}{\epsilon(k)} (x_N^*)^{\top} Q_N x_N^* \quad \text{(B.46)}$$

for some $(x(k), p_1(k))$ with corresponding optimal input and state sequences

$$u^*(x(k), p_1(k)) = \{u_0^*, \ldots, u_{N-1}^*\},$$
$$x^*(x(k), p_1(k)) = \{x_0^*, \ldots, x_N^*\}. \quad \text{(B.47)}$$

Furthermore, let the suboptimal value function $\tilde{V}_1 : \mathbb{R}^n \times \mathbb{N} \times \mathbb{R}_+ \to \mathbb{R}_+$ at sampling instant $k + 1$ be denoted as

$$
\begin{aligned}
\tilde{V}_1(x(k+1), p_1(k+1)) = &\sum_{i=0}^{N-1} (x_i^+)^\top Q x_i^+ + (u_i^+)^\top R u_i^+ \\
&+ \frac{\Gamma(x(k+1))}{\epsilon(k+1)} (x_N^+)^\top Q_N x_N^+
\end{aligned}
\tag{B.48}
$$

for some $(x(k+1), p_1(k+1))$ with corresponding suboptimal input and state sequences

$$
\begin{aligned}
u^+(x(k+1), p_1(k+1)) &= \{u_0^+, \dots, u_{N-2}^+, u_{N-1}^+\} = \{u_1^*, \dots, u_{N-1}^*, 0\} \\
x^+(x(k+1), p_1(k+1)) &= \{x_0^+, \dots, x_{N-1}^+, x_N^+\} \\
&= \{x_1^* + e_0(k), \dots, x_N^* + e_{N-1}(k), \\
&\quad\quad A_1(k+1) x_N^* + e_N(k)\}
\end{aligned}
\tag{B.49}
$$

where the error sequence $\{e_i(k)\}_{i=0}^N$ is defined as

$$
\begin{aligned}
P_{i+1}(k, x(k), u_0^*, \dots, u_i^*) + e_i(k) &= P_i(k+1, x(k+1), u_0^+, \dots, u_{i-1}^+), \ i = 1, \dots, N, \\
P_{i+1}(k, x(k), u_0^*, \dots, u_i^*) + e_i(k) &= x(k+1), \ i = 0.
\end{aligned}
\tag{B.50}
$$

In the following, we use the optimal value function $V_1^*(x(k), p_1(k))$ as Lyapunov function and consider the difference between two consecutive sampling instants. More precisely, we have

$$
\begin{aligned}
&V_1^*(x(k+1), p_1(k+1)) - V_1^*(x(k), p_1(k)) \\
&\leq \tilde{V}_1(x(k+1), p_1(k+1)) - V_1^*(x(k), p_1(k)) \\
&= \sum_{i=1}^{N} (x_i^* + e_i(k))^\top Q (x_i^* + e_i(k)) + (u_i^*)^\top R u_i^* \\
&\quad + \frac{\Gamma(x(k+1))}{\epsilon(k+1)} (A_1(k+1) x_N^* + e_N(k))^\top Q_N (A_1(k+1) x_N^* + e_N(k)) \\
&\quad - \sum_{i=0}^{N-1} (x_i^*)^\top Q x_i^* + (u_i^*)^\top R u_i^* - \frac{\Gamma(x(k))}{\epsilon(k)} (x_N^*)^\top Q_N x_N^* \\
&= -(x_0^*)^\top Q x_0^* - (u_0^*)^\top R u_0^* - \frac{\Gamma(x(k))}{\epsilon(k)} (x_N^*)^\top Q_N x_N^* \\
&\quad + (x_N^* + e_{N-1}(k))^\top Q_N (x_N^* + e_{N-1}(k)) + \frac{\Gamma(x(k+1))}{\epsilon(k+1)} (A_1(k+1) x_N^* \\
&\quad + e_N(k))^\top Q_N (A_1(k+1) x_N^* + e_N(k)) + r(k)
\end{aligned}
\tag{B.51}
$$

with $r(k) = \sum\limits_{i=1}^{N-1}(x_i^* + e_i(k))^\top Q(x_i^* + e_i(k)) - (x_i^*)^\top Q x_i^*$. From the maximal eigenvalue inequality together with Young's inequality, we know that $2x^\top Q y$ with $Q > 0$ can be upper bounded by

$$2x^\top Q y \le 2\lambda_{\max}(Q)x^\top y \le c_1(x^\top x + y^\top y) \tag{B.52}$$

with $c_1 = \lambda_{\max}(Q) > 0$ for all x, y. Hence, we can conclude that

$$(x_N^* + e_{N-1}(k))^\top Q_N(x_N^* + e_{N-1}(k)) \le c_2(x_N^*)^\top x_N^* \\ + c_3(e_{N-1}(k))^\top e_{N-1}(k) \tag{B.53}$$

with $c_2 = c_3 = 2\lambda_{\max}(Q_N) > 0$. Moreover, following the ideas of (B.53), we get

$$(A_1(k+1)x_N^* + e_N(k))^\top Q_N(A_1(k+1)x_N^* + e_N(k)) \le c_6(x_N^*)^\top x_N^* \\ + c_5 e_N(k)^\top e_N(k) \tag{B.54}$$

with $c_6 = 2\lambda_{\max}(Q_N)\lambda_{\max}(A_1(k+1)^\top Q_N A_1(k+1)) \ge 0$ and $c_5 = 2\lambda_{\max}(Q_N) > 0$. Inserting (B.53) and (B.54) into eq. (B.51) and using the fact that $-\frac{\Gamma(x(k))}{\epsilon(k)}(x_N^*)^\top Q_N x_N^* \le 0$ for $\Gamma(x(k)) = \alpha x(k)^\top x(k), \alpha > 0$, we have

$$\tilde{V}_1(x(k+1), p_1(k+1)) - V_1^*(x(k), p_1(k)) \tag{B.55}$$

$$\le -(x_0^*)^\top Q x_0^* - (u_0^*)^\top R u_0^* + c_2(x_N^*)^\top x_N^* + \frac{\Gamma(x(k+1))}{\epsilon(k+1)}c_6(x_N^*)^\top x_N^* + \tilde{r}(k) \tag{B.56}$$

with $\tilde{r}(k) = r(k) + c_3(e_{N-1}(k))^\top e_{N-1}(k) + \frac{\Gamma(x(k+1))}{\epsilon(k+1)}c_5(e_N(k))^\top e_N(k)$. We now prove that $\frac{\Gamma(x(k+1))}{\epsilon(k+1)} \le c$ for some $c > 0$. From Theorem 5, we know that $x_N^*(x, p_1) = K_{1,N}(x, p_1)x$ for all (x, p_1) which can be rewritten into $x_N^*(x, p_1)^\top x_N^*(x, p_1) = x^\top K_{1,N}(x, p_1)^\top K_{1,N}(x, p_1)x$. We can assume without the loss of generality that $x^\top K_{1,N}(x, p_1)^\top K_{1,N}(x, p_1)x > 0$ as $x^\top K_{1,N}(x, p_1)^\top K_{1,N}(x, p_1)x = 0$ refers to the trivial case $x_N^* = 0$. Hence, we result in

$$c_7 x^\top x \le x_N^*(x, p_1)^\top x_N^*(x, p_1) \tag{B.57}$$

with $c_7 = \lambda_{\min}(K_{1,N}(x, p_1)^\top K_{1,N}(x, p_1)) > 0$. Due to $c_7 > 0$, we get

$$x^\top x \le \frac{1}{c_7}x_N^*(x, p_1)^\top x_N^*(x, p_1) \tag{B.58}$$

for all (x, p_1). Now, consider $x = x(k+1)$ and $p_1 = p_1(k+1)$. Furthermore, inserting (B.58) into $\frac{\Gamma(x(k+1))}{\epsilon(k+1)}$ with $\Gamma(x(k+1)) = \alpha x(k+1)^\top x(k+1)$, $\alpha > 0$ yields to

$$
\begin{aligned}
\frac{\Gamma(x(k+1))}{\epsilon(k+1)} &= \frac{\alpha x(k+1)^\top x(k+1)}{\epsilon(k+1)} \\
&\leq \frac{\alpha}{c_7} \frac{x_N^*(x(k+1), p_1(k+1))^\top x_N^*(x(k+1), p_1(k+1))}{\epsilon(k+1)} \\
&= c_8 \frac{x_N^*(x(k+1), p_1(k+1))^\top x_N^*(x(k+1), p_1(k+1))}{\epsilon(k+1)}
\end{aligned}
\tag{B.59}
$$

with $c_8 = \frac{\alpha}{c_7} > 0$. By inserting eq. (3.11) of Lemma 7 into (B.59), we can conclude that

$$
\frac{\Gamma(x(k+1))}{\epsilon(k+1)} \leq c_9 \frac{V_2^*(x(k+1), p_2(k+1)) + \epsilon(k+1)\rho}{\epsilon(k+1)}
\tag{B.60}
$$

with $c_9 = \frac{c_8}{\lambda_{\max}(Q_N)} > 0$. Due to optimality and Assumption 15 c.), we get that $V_2^*(x(k+1), p_2(k+1)) \leq c_5\beta(k+1)$ with $c_5 > 0$ which directly leads to

$$
\frac{\Gamma(x(k+1))}{\epsilon(k+1)} \leq c_9 \frac{c_5\beta(k+1) + \epsilon(k+1)\rho}{\epsilon(k+1)}.
\tag{B.61}
$$

Due to $\epsilon(k) > \beta(k)$ for all $k \in \mathbb{N}$, we have that

$$
\frac{\Gamma(x(k+1))}{\epsilon(k+1)} \leq c_{10}
\tag{B.62}
$$

with $c_{10} = c_9 c_5 c + c_9 \rho > 0$. Moreover, from eq. (3.11) of Lemma 7, we have that

$$
x_N^*(x, p_1)^\top x_N^*(x, p_1) \leq c_{11}(V_2^*(x, p_1) + \epsilon\rho)
\tag{B.63}
$$

with $c_{11} = \frac{1}{\lambda_{\min}(Q_N)} > 0$. Inserting (B.63), (B.62) into (B.56), we finally get

$$
\begin{aligned}
\tilde{V}_1(x(k+1), &p_1(k+1)) - V_1^*(x(k), p_1(k)) \\
\leq &-(x_0^*)^\top Q x_0^* - (u_0^*)^\top R u_0^* \\
&+ c_2 c_{11}(V_2(x(k), p_1(k))) + c_{10} c_6 c_{11}(V_2(x(k), p_1(k)) + \epsilon(k)\rho) + r^*(k)
\end{aligned}
\tag{B.64}
$$

with $r^*(k) = r(k) + c_3(e_{N-1}(k))^\top e_{N-1}(k) + c_{10}c_5 e_N(k)^\top e_N(k)$. If we now define the disturbance $d(k)$ at sampling instant k as

$$d(k) = c_2 c_{11}(V_2(x(k), p_1(k)) + \epsilon(k)\rho) + c_{10} c_6 c_{11}(V_2(x(k), p_1(k)) + \epsilon(k)\rho) + r^*(k).$$

$$(B.65)$$

We know, due to Definition 11 and Lemma 12, that the closed-loop (3.3) and (3.2) is ISS. Moreover, a direct consequence of Assumption 15 c.) is that $\lim_{k\to\infty} V_2(x(k), p_2(k)) = 0$. Furthermore, it holds that $\lim_{k\to\infty} \epsilon(k) = 0$. Additionally, due to Assumption 15 a.) and

$$P_{i+1}(k, x(k), u_0^*, \dots, u_i^*) + e_i(k) = P_i(k+1, x(k+1), u_0^*, \dots, u_{i-1}^*),\ i = 1, \dots, N,$$
$$P_{i+1}(k, x(k), u_0^*, \dots, u_i^*) + e_i(k) = x(k+1),\ i = 0$$

$$(B.66)$$

can be written as

$$A_{i+1}(k)x(k) + \sum_{l=0}^{i} B_{i-l}(k)u_l^* + e_i(k) = A_i(k+1)x(k+1) + \sum_{l=0}^{i-1} B_{i-1-l}(k+1)u_l^*$$

$$(B.67)$$

for $i = 1, \dots, N$ and

$$A_{i+1}(k)x(k) + \sum_{l=0}^{i} B_{i-l}(k)u_l^* + e_i(k) = x(k+1) \qquad (B.68)$$

for $i = 0$. Furthermore, due to Assumption 15 b.), we know that $\lim_{k\to\infty} e_i(k) = 0$ for $i = 0, \dots, N$. Hence, we can conclude that $\lim_{k\to\infty} d(k) = 0$. Together with the ISS property of the closed-loop, we can conclude that for any initial state $x(0)$, the state of the closed-loop $x(k)$ will converge to zero under the stated assumptions. $\qquad \square$

B.12 Proof of Lemma 8

From the first-order optimality condition for the right-hand side of (3.18), we get

$$\nabla_x f(x_{k+1}^*, k) = -\nabla_x D(x_{k+1}^*, x_k^*). \qquad (B.69)$$

Moreover, due to the convexity of f in the first argument, it holds for all $x \in \mathbb{R}^n$

$$f(x,k) \geq f(x_{k+1}^*,k) + \nabla_x f(x_{k+1}^*,k)^\top (x - x_{k+1}^*). \tag{B.70}$$

Evaluating (B.70) at a time invariant minimizer $x^* \in X$, inserting (B.69) into (B.70) and subtracting $f(x_{k+1}^*,k)$, we get

$$f(x^*,k) - f(x_{k+1}^*,k) \geq -\nabla_x D(x_{k+1}^*,x_k^*)^\top (x^* - x_{k+1}^*). \tag{B.71}$$

Inserting the definition of the Bregman distance and the gradient of the Bregman distance with $\nabla_x D(x,y) = \nabla g(x) - \nabla g(y)$ into (B.71) yields

$$\begin{aligned} f(x^*,k) - f(x_{k+1}^*,k) &\geq [\nabla g(x_k^*) - \nabla g(x_{k+1}^*)]^\top (x^* - x_{k+1}^*) \\ &= D(x^*,x_{k+1}^*) + D(x_{k+1}^*,x_k^*) - D(x^*,x_k^*), \end{aligned} \tag{B.72}$$

where the last equality follows from the three-point identity of the Bregman distance, see also (Y. Censor and S. Zenios, 1991, Proof of Proposition 3.6). Rearranging eq. (B.72), yields to

$$D(x^*,x_{k+1}^*) \leq D(x^*,x_k^*) - D(x_{k+1}^*,x_k^*) + f(x^*,k) - f(x_{k+1}^*,k), \tag{B.73}$$

and hence

$$\begin{aligned} D(x^*,x_{k+1}^*) &- D(x^*,x_k^*) \\ &\leq -D(x_{k+1}^*,x_k^*) + f(x^*,k) - f(x_{k+1}^*,k). \end{aligned} \tag{B.74}$$

The following steps consist of two parts. In i.), we prove that $\lim_{k\to\infty} f(x_k^*,k) = \lim_{k\to\infty} f(x^*,k)$. In ii.), we prove that x_k^* converges to a single point. Combining both results leads to the convergence of x_k^* to a single minimizer of f.

i.) Due to D being non-negative, eq. (B.74) transforms into

$$f(x_{k+1}^*,k) - f(x^*,k) \leq D(x^*,x_k^*) - D(x^*,x_{k+1}^*). \tag{B.75}$$

Summing both sides for $k = 0,\ldots,\infty$, and, due to the fact of the right hand side being a telescopic sum, we result in

$$\begin{aligned} \sum_{k=0}^{\infty} f(x_{k+1}^*,k) - f(x^*,k) &\leq \sum_{k=0}^{\infty} D(x^*,x_k^*) - D(x^*,x_{k+1}^*) \\ &\leq D(x^*,x_0^*) - D(x^*,x_\infty^*) \end{aligned} \tag{B.76}$$

Due to optimality, we know that $f(x_k^*,k) - f(x^*,k) > 0$ for $x_k^* \notin \mathcal{X}$ for all $k \in \mathbb{N}$. Moreover, we know that D being non-negative. Hence, (B.76) transforms into

$$0 \leq \sum_{k=0}^{\infty} [f(x_k^*,k) - f(x^*,k)] \leq D(x^*,x_0^*) \tag{B.77}$$

Since D is non-negative and g continuous differentiable, we know that there exist a $c > 0$ such that $D(x^*, x_0^*) \leq c$. Furthermore, we know that $f(x_k^*, k) - f(x^*, k)$ is strictly positive if $x_k^* \notin \mathcal{X}$. Hence, we infer that $\lim_{k\to\infty} f(x_k^*, k) = f^* = 0$ where $f(x^*, k) = f^* = 0$ for all $k \in \mathbb{N}$.

ii.) In the following, we like to prove that x_k^* converges to a single minimizer of f. From (B.75), we know that

$$D(x^*, x_{k+1}^*) - D(x^*, x_k^*) \leq 0 \tag{B.78}$$

for all $x^* \in \mathcal{X}$. Thus, we know that $D(x^*, x_k^*)$ is non-increasing for all $x^* \in \mathcal{X}$. Due to the $D(x, y) > 0$ for all $x \neq y$, $D(x, y) = 0$ for $x = y$ and D being convex in both arguments, we know that D is radially unbounded, see A. Ahmadi and R. Jungers (2018)[Lemma 4.1]. Hence, due to $D(x^*, x_k^*)$ being non-increasing, the fact that $D(x, y)$ is non-negative and its radial unboundedness, we can conclude that x_k^* is bounded. As x_k^* is bounded, there exists a subsequence $\{x_{k_j}^*\}$ with limit point z such that $\lim_{k_j\to\infty} x_{k_j}^* = z$. Since $\lim_{k\to\infty} f(x_k^*, k) = f^*$, it holds that $\lim_{k_j\to\infty} f(x_{k_j}^*, k_j) = f^*$ and thus we have that $z \in \mathcal{X}$. Due to (B.78), we know that $D(z, x_k^*)$ is non-increasing. Hence, for any $k \geq k_j$, we have that

$$D(z, x_k^*) \leq D(z, x_{k_j}^*). \tag{B.79}$$

Due to continuity, see Protter and Morrey (2012)[Theorem 2.7] of D and the fact that $\lim_{k_j\to\infty} x_{k_j}^* = z$, we know that $\lim_{k_j\to\infty} D(z, x_{k_j}^*) = 0$. Hence, we result, due to (B.79), into

$$\lim_{k\to\infty} D(z, \tilde{x}_k) = 0. \tag{B.80}$$

Assume, we have another subsequence $x_{k_l}^*$ with a different limit point $\bar{x}$ such that $\lim_{k_l\to\infty} x_{k_l}^* = \bar{x}$. Since (B.80) is true, we know that (B.80) holds for any subsequence of $\{x_k^*\}$. Thus, it holds that $\lim_{k_l\to\infty} D(z, x_{k_l}^*) = 0$ which means that $\lim_{k_l\to\infty} x_{k_l}^* = z = \bar{x}$.

$\square$

B.13 Proof of Theorem 7

(i) By Assumptions 17 and 18 and by eliminating the variable e in (3.14), we can apply Lemma 8. Hence, we have $\lim_{k\to\infty} \theta(k) = \theta^*$ where $\theta(k)$ is bounded for all $k \in \mathbb{N}$.

(ii) From the construction (3.19) of the predictor maps and the fact that $s(k) - R(k)\theta^*(k) = e^*(k)$ corresponds to the linear system of equations

$$x(j) = A(k)x(j-1) + B(k)u(j-1) + e^*_{j-k+\bar{N}}(k), \tag{B.81}$$

$j = k...k - \bar{N} + 1$, $e^*(k) = [e^*_{\bar{N}}(k)^\top \ldots e^*_1(k)^\top]$, as well as from the convergence property (i), i.e. by $\lim_{k\to\infty} A(k) = \hat{A}$, $\lim_{k\to\infty} B(k) = \hat{B}$, property (3.6), (3.7) in Assumption 15 is satisfied. It remains to show property (3.8). Notice that by (B.81), we have for any $k \in \mathbb{N}$ and $j = 0...\bar{N} - 1$

$$x(j + k - \bar{N} + 1) = A_j(k)x(k - \bar{N} + 1)$$

$$+ \sum_{l=0}^{j-1} B_{j-1-l}(k)u(l + k - \bar{N} + 1) + \sum_{l=0}^{j-1} A(k)^{j-1-l}e^*_{l+1}(k). \tag{B.82}$$

Since $\lim_{k\to\infty} A_i(k) = \hat{A}_i$, $\lim_{k\to\infty} B_i(k) = \hat{B}_i$, we have

$$x(j + k - \bar{N} + 1) = \hat{A}_j x(k - \bar{N} + 1)$$

$$+ \sum_{l=0}^{j-1} \hat{B}_{j-1-l}u(l + k - \bar{N} + 1) + \sum_{l=0}^{j-1} A(k)^{j-1-l}e^*_{l+1}(k)$$

$$+ (A_j(k) - \hat{A}_j)x(k - \bar{N} + 1) \tag{B.83}$$

$$+ \sum_{l=0}^{j-1} (B_{j-1-l}(k) - \hat{B}_{j-l-1})u(l + k - \bar{N} + 1).$$

By Lemma 8, $\lim_{k\to\infty} e^*_{l+1}(k) = 0$, $l = 0...\bar{N} - 1$, hence there exists a sequence $\{\omega_1(k)\}_{k\in\mathbb{N}}$ that converges to zero and such that for all $j = 0...\bar{N} - 1$ it holds: $\|\sum_{l=0}^{j-1} A(k)^{j-1-l}e^*_{l+1}(k)\|^2 \leq \omega_1(k)$. By Young's inequality, there exist sequences $\{\omega_2(k)\}_{k\in\mathbb{N}}$, $\{\omega_3(k)\}_{k\in\mathbb{N}}$ that both converge to zero and such that $\|(A_j(k) - \hat{A}_j)x(k - \bar{N} + 1)\|^2 \leq \omega_2(k)\|x(k - \bar{N} + 1)\|^2$ and such that for all $j = 0...\bar{N} - 1$ $\|\sum_{l=0}^{j-1}(B_{j-1-l}(k) - \hat{B}_{j-l-1})u(l + k - \bar{N} + 1)\|^2 \leq \omega_3(k)\sum_{l=0}^{j-1}\|u(l + k - \bar{N} + 1)\|^2$. Hence

$$x(j + k - \bar{N} + 1) = \hat{A}_j x(k - \bar{N} + 1) + e_j(k - \bar{N} + 1)$$

$$+ \sum_{l=0}^{j-1} \hat{B}_{j-1-l}u(l + k - \bar{N} + 1), \tag{B.84}$$

where $e_j(k - \bar{N} + 1) = \sum_{l=0}^{j-1} A(k)^{j-1-l}e^*_{l+1}(k) + (A_j(k) - \hat{A}_j)x(k - \bar{N} + 1) + \sum_{l=0}^{j-1}(B_{j-1-l}(k) - \hat{B}_{j-l-1})u(l + k - \bar{N} + 1)$, $j = 0...\bar{N} - 1$, satisfies the error bounds in Assumption 15 b.) as shown above. In more detail, since (B.84) holds for any $k \in \mathbb{N}$, let $k = \tilde{k} + \tilde{N} - 1$, then we get the desired asymptotically correct prediction property (3.8) with respect to (3.17) with $N = \tilde{N} - 1$:

$$
\hat{A}_j x(\tilde{k}) + \sum_{l=0}^{j-1} \hat{B}_{j-1-l} u(l + \tilde{k}) + e_j(\tilde{k})
$$
$$
= x(j + \tilde{k}) = A^j x(\tilde{k}) + \sum_{l=0}^{j-1} A^{j-1-l} B u(l + \tilde{k}).
$$

(B.85)

$\square$

B.14 Proof of Lemma 9

We use the Kalman decomposition. Because we consider only input and output data and because Assumption 13 holds, we can assume without loss of generality that (F, G, H) in (3.1) is structured as follows

$$
\begin{bmatrix} z_1(k+1) \\ z_2(k+1) \\ z_3(k+1) \end{bmatrix} = \begin{bmatrix} F_1 & F_2 & 0 \\ 0 & F_3 & 0 \\ F_4 & F_5 & F_6 \end{bmatrix} \begin{bmatrix} z_1(k) \\ z_2(k) \\ z_3(k) \end{bmatrix} + \begin{bmatrix} G_1 \\ 0 \\ G_2 \end{bmatrix} v(k)
$$
$$
y(k) = \begin{bmatrix} H_1 & H_2 & 0 \end{bmatrix} \begin{bmatrix} z_1(k) \\ z_2(k) \\ z_3(k) \end{bmatrix},
$$

(B.86)

where F_3, F_6 are stable (eigenvalues are in the interior of the complex unit disc) and the subsystem

$$
\begin{bmatrix} z_1(k+1) \\ z_2(k+1) \end{bmatrix} = \begin{bmatrix} F_1 & F_2 \\ 0 & F_3 \end{bmatrix} \begin{bmatrix} z_1(k) \\ z_2(k) \end{bmatrix} + \begin{bmatrix} G_1 \\ 0 \end{bmatrix} v(k)
$$
$$
:= F_s z_s(k) + G_s v(k)
$$
$$
y(k) = \begin{bmatrix} H_1 & H_2 \end{bmatrix} z_s(k) := H_s z_s(k)
$$

(B.87)

is observable and stabilizable. Owing to the cascaded structure, it follows that if the state and input of the subsystem (B.87) goes to zero, then also the state (and the input) of the overall system (B.86), since F_6 is stable. Note further that the output sequences of the subsystem are equivalent to the output sequences of the overall system (for the same input sequences). Since the subsystem is observable and $m \geq n$ is known, it follows that if $x(k) = \phi_y(y(k), ..., y(k - $

$m + 1)) = \left[y(k)^{\top}, ..., y(k - m + 1)^{\top}\right]^{\top}$ and $u(k) = \phi_{\mathrm{v}}(v(k), ..., v(k - m + 1)) = \left[v(k)^{\top}, ..., v(k - m + 1)^{\top}\right]^{\top}$ goes to zero, then also $\{v(k)\}_{k \in \mathbb{N}}$, $\{y(k)\}_{k \in \mathbb{N}}$ of the subsystem (B.87) (as well as of the overall system (B.86)). As a final step, consider $x(k) = \phi_{\mathrm{y}}(y(k), ..., y(k - m + 1)) = \left[y(k)^{\top}, ..., y(k - m + 1)^{\top}\right]^{\top}$, which is given by

$$\begin{bmatrix} H_s F_s^{m-1} z_s(k - m + 1) + H_s \sum_{l=0}^{m-2} F_s^{m-2-l} G_s v(k - m + 1 + l) \\ \vdots \\ H_s F_s z_s(k - m + 1) + H_s G_s v(k - m + 1) \\ H_s z_s(k - m + 1) \end{bmatrix}. \tag{B.88}$$

Due to (3.20), (B.88) can be compactly written as $x(k) = O z_s(k - m + 1) + R \bar{u}(k - 1)$ with $\bar{u}(k - 1) = \left[v(k - 1)^{\top}, ..., v(k - m + 1)^{\top}\right]^{\top}$. Since (B.87) is observable, the (observability) matrix O has full column rank, hence we have $z_s(k - m + 1) = (O^{\top} O)^{-1} O^{\top} x(k) - (O^{\top} O)^{-1} O^{\top} R \bar{u}(k - 1)$. Thus

$$\begin{aligned} x(k + 1) &= O z_s(k - m + 2) + R \bar{u}(k) \\ &= O F_s z_s(k - m + 1) + O G_s v(k - m + 1) + R \bar{u}(k) \\ &= O F_s ((O^{\top} O)^{-1} O^{\top} x(k) - (O^{\top} O)^{-1} O^{\top} R \bar{u}(k - 1)) \\ &\quad + O G_s v(k - m + 1) + R \bar{u}(k), \end{aligned} \tag{B.89}$$

Due to $u(k) = \left[v(k)^{\top}, ..., v(k - m + 1)^{\top}\right]^{\top}$, eq. (B.89) can be written as $x(k + 1) = A x(k) + B u(k)$. Consequently $\{u(k)\}_{k \in \mathbb{N}}$, $\{x(k)\}_{k \in \mathbb{N}}$ satisfy Assumption 18. $\qquad \square$

C
Auxiliary results

C.1 Vector-matrix formulation: Trajectory tracking rbMPC algorithm

In the following, we will derive the cost matrices and the inequality matrices of (4.20) and (4.21). By introducing the stacked help variables $\zeta = [\zeta_0^\top \ \zeta_1^\top \ \ldots \ \zeta_N^\top]^\top \in \mathbb{R}^{(N+1)n}$, $U = [u_0^\top \ \ldots \ u_{N-1}^\top]^\top \in \mathbb{R}^{Nm}$, $U_{-1} = [u_{-1}^\top \ u_{-1}^\top \ \ldots \ u_{-1}^\top]^\top \in \mathbb{R}^{Nm}$ and $E = [e_0^\top \ e_1^\top \ e_{N-1}^\top] \in \mathbb{R}^{Nn}$, we can rewrite the system dynamics in the open-loop optimal control problem (4.19) as

$$\zeta = \Omega_1 \zeta_0 + \Omega_2 U + \Omega_3 E \tag{C.1}$$

with the auxiliary matrices $\Omega_1 \in \mathbb{R}^{Nn \times n}$, $\Omega_2 \in \mathbb{R}^{Nn \times Nm}$, and $\Omega_3 \in \mathbb{R}^{Nn \times Nn}$ defined as

$$\Omega_1 = \begin{bmatrix} \prod_{i=1}^{1} A_i \\ \prod_{i=1}^{2} A_i \\ \vdots \\ \prod_{i=1}^{N} A_i \end{bmatrix}, \ \Omega_2 = \begin{bmatrix} B_1 & 0 & \ldots & 0 \\ A_2 B_1 & B_2 & \ldots & 0 \\ \vdots & \vdots & \ddots & \vdots \\ \prod_{i=2}^{N} A_i B_1 & \ldots & \ldots & B_N \end{bmatrix},$$

$$\Omega_3 = \begin{bmatrix} I_n & 0 & \ldots & & \ldots & 0 \\ A_2 & I_n & 0 & & \ldots & 0 \\ \vdots & & \ddots & & \vdots & \vdots \\ \vdots & & \ldots & & \ddots & \vdots \\ \prod_{i=2}^{N} A_i & \ldots & & \ldots & A_{N-1} & I_n \end{bmatrix}.$$

For the sake of clarity, we neglect the time dependence of the matrices $A_i(k)$ and $B_i(k)$ and just write A_i and B_i. Moreover, we can rewrite $u_i = u_{i-1} + \Delta u_i$ for $i = 0, \ldots, N-1$ in vector-matrix form and thus get

$$U = \Gamma_1 \Delta U + U_{-1} \tag{C.2}$$

where the auxiliary matrix $\Gamma_1 \in \mathbb{R}^{Nm \times Nm}$ is defined as

$$\Gamma_1 = \begin{bmatrix} I_m & O_m & \dots & O_m \\ I_m & I_m & \dots & O_m \\ I_m & \dots & \dots & I_m \end{bmatrix}.$$

Together with the stacked state reference $\zeta_{\text{ref,s}}(k) = [\zeta_{\text{ref}}(k+1)^\top \dots \zeta_{\text{ref}}(k+N)^\top]^\top \in \mathbb{R}^{Nn}$ we can conclude that

$$
\begin{aligned}
H &= \Gamma_1^\top \Omega_2^\top \tilde{Q} \Omega_2 \Gamma_1 + \tilde{R} \\
F &= 2[(\Omega_1 \zeta_0 + \Omega_2 U_{-1} + \Omega_3 E)^\top \tilde{Q} \Omega_2 \Gamma_1 - \zeta_{\text{ref,s}}(k)^\top \tilde{Q} \Omega_2 \Gamma_1] \\
s &= (\Omega_1 \zeta_0 + \Omega_2 U_{-1} + \Omega_3 E - \zeta_{\text{ref,s}}(k))^\top \tilde{Q}(\Omega_1 \zeta_0 + \Omega_2 U_{-1} + \Omega_3 E - \zeta_{\text{ref,s}}(k)) \\
&\quad + (\zeta_0 - \zeta_{\text{ref,0}}(k))^\top Q(\zeta_0 - \zeta_{\text{ref,0}}(k))
\end{aligned}
$$

$$\text{(C.3)}$$

with the help matrices

$$\tilde{Q} = \begin{bmatrix} I_{N-1} \otimes Q & 0 \\ 0 & P \end{bmatrix}, \quad R = \begin{bmatrix} I_N \otimes R \end{bmatrix}. \tag{C.4}$$

Now, we want to define the inequality matrix G and time-varying vector $d(k)$ where the stacked input reference is denoted by $u_{\text{ref,s}}(k) = [u_{\text{ref}}(k)^\top \dots u_{\text{ref}}(k+N-1)^\top]^\top \in \mathbb{R}^{Nm}$. The state, input, and Δ input constraints can be written in vector-matrix form. More precisely, we have

$$\tilde{C}_x(\zeta - \zeta_{\text{ref,s}}(k)) \le \tilde{d}_x(k) \tag{C.5}$$

$$\tilde{C}_u(u - u_{\text{ref,s}}(k)) \le \tilde{d}_u(k) \tag{C.6}$$

$$\tilde{C}_{\Delta u} \Delta u \le \tilde{d}_{\Delta u} \tag{C.7}$$

with $\tilde{d}_x(k) = \tilde{d}_x - \tilde{C}_x \zeta_{\text{ref,s}}(k)$ and $\tilde{d}_u(k) = \tilde{d}_u - \tilde{C}_u u_{\text{ref,s}}(k)$ where

$$\tilde{C}_x = \begin{bmatrix} C_x & \dots & 0 \\ \vdots & \ddots & \vdots \\ 0 & \dots & C_x \end{bmatrix}, \tilde{d}_x = \begin{bmatrix} d_x \\ \vdots \\ d_x \end{bmatrix}, \tilde{C}_u = \begin{bmatrix} C_u & \dots & 0 \\ \vdots & \ddots & \vdots \\ 0 & \dots & C_u \end{bmatrix}, \tilde{d}_u = \begin{bmatrix} d_u \\ \vdots \\ d_u \end{bmatrix} \text{ and}$$

$$\text{(C.8)}$$

$$\tilde{C}_{\Delta u} = \begin{bmatrix} C_{\Delta u} & \dots & 0 \\ \vdots & \ddots & \vdots \\ 0 & \dots & C_{\Delta u} \end{bmatrix}, \tilde{d}_{\Delta u} = \begin{bmatrix} d_{\Delta u} \\ \vdots \\ d_{\Delta u} \end{bmatrix}. \tag{C.9}$$

With the help of eq. (C.1) and (C.2), eq. (C.5) - (C.7) can be transformed into

$$\tilde{C}_x \Omega_2 \Gamma_1 \Delta U + r_x \leq \tilde{d}_x(k) \tag{C.10}$$

$$\tilde{C}_u \Gamma_1 \Delta U + r_u \leq \tilde{d}_u(k) \tag{C.11}$$

$$\tilde{C}_{\Delta u} \Delta u \leq \tilde{d}_{\Delta u} \tag{C.12}$$

with $r_x = \tilde{C}_x(\Omega_1 \zeta_0 + \Omega_2 U_{-1} + \Gamma_3 E - \zeta_{\text{ref,s}}(k))$ and $r_u = \tilde{C}_u(U_{-1} - u_{\text{ref,s}}(k))$. Thus, we have that

$$G = \begin{bmatrix} C_x \Omega_2 \Gamma_1 \\ \tilde{C}_u \Gamma_1 \\ C_{\Delta u} \end{bmatrix}, \ r = \begin{bmatrix} r_x \\ r_u \\ 0 \end{bmatrix} \text{ and } d(k) = \begin{bmatrix} \tilde{d}_x(k) \\ \tilde{d}_u(k) \\ \tilde{d}_{\Delta u} \end{bmatrix}. \tag{C.13}$$

C.2 Vector-matrix formulation: Path following rbMPC algorithm

In the following, we will derive the cost matrices and the inequality matrices from (4.27) and (4.28), respectively. By introducing the stacked help variables $\zeta = [\zeta_0^\top \ \zeta_1^\top \ \dots \ \zeta_N^\top]^\top \in \mathbb{R}^{(N+1)n}$, $U = [u_0^\top \ \dots \ u_{N-1}^\top]^\top \in \mathbb{R}^{Nm}$, $U_{-1} = [u_{-1}^\top \ u_{-1}^\top \ \dots \ u_{-1}^\top]^\top \in \mathbb{R}^{Nm}$ and $E = [e_0^\top \ e_1^\top \ \dots \ e_{N-1}^\top] \in \mathbb{R}^{Nn}$, we can rewrite the system dynamics in the open-loop optimal control problem (4.25) into

$$\zeta = \Omega_1 \zeta_0 + \Omega_2 U + \Omega_3 E \tag{C.14}$$

with the auxiliary matrices $\Omega_1 \in \mathbb{R}^{Nn \times n}$, $\Omega_2 \in \mathbb{R}^{Nn \times Nm}$ and $\Omega_3 \in \mathbb{R}^{Nn \times Nn}$ being denoted by

$$\Omega_1 = \begin{bmatrix} \prod_{i=1}^{1} A_i \\ \prod_{i=1}^{2} A_i \\ \vdots \\ \prod_{i=1}^{N} A_i \end{bmatrix}, \ \Omega_2 = \begin{bmatrix} B_1 & 0 & \dots & 0 \\ A_2 B_1 & B_2 & \dots & 0 \\ \vdots & \vdots & \ddots & \vdots \\ \prod_{i=2}^{N} A_i B_1 & \dots & \dots & B_N \end{bmatrix},$$

$$\Omega_3 = \begin{bmatrix} I_n & 0 & \dots & \dots & 0 \\ A_2 & I_n & 0 & \dots & 0 \\ \vdots & \dots & \ddots & \vdots & \vdots \\ \vdots & \dots & & \ddots & \vdots \\ \prod_{i=2}^{N} A_i & \dots & \dots & A_{N-1} & I_n \end{bmatrix}.$$

Due to space limitations, we write A_i, B_i instead of $A(\theta_i)$ and $B(\theta_i)$. Moreover, we can rewrite $u_i = u_{i-1} + \Delta u_i$ for $i = 0, \ldots, N-1$ in vector-matrix form

$$U = \Gamma_1 \Delta U + U_{-1}, \tag{C.15}$$

where $\Gamma_1 \in \mathbb{R}^{Nm \times Nm}$ and $U_{-1} \in \mathbb{R}^{Nm}$ are defined according to eq. (C.2) and together with $\Delta U = \Pi_2 w$ and $\Pi_2 = [I_{Nm \times Nm} \ 0_{Nm \times N}] \in \mathbb{R}^{Nm \times N(m+1)}$, we have that

$$U = \Gamma_1 \Pi_2 w + U_{-1}. \tag{C.16}$$

Combining eq. (C.14) and (C.16) yields to

$$\xi = \Omega_1 \xi_0 + \Omega_2 \Gamma_1 \Pi_2 w + \Omega_2 U_{-1} + \Omega_3 E. \tag{C.17}$$

The stacked reference $\xi_{\text{ref}} = \left[\xi_{\text{ref}}(\theta_1)^\top \ \cdots \ \xi_{\text{ref}}(\theta_N)^\top \right]^\top \in \mathbb{R}^{Nn}$ can be written as

$$\xi_{\text{ref}} = \Lambda_1 + \Lambda_2 \Lambda_3 + \Lambda_2 \Gamma_2 \Pi_3 w, \tag{C.18}$$

where the auxiliary matrices $\Lambda_1 \in \mathbb{R}^{Nn}$, $\Lambda_2 \in \mathbb{R}^{Nn \times N}$, $\Gamma_2 \in \mathbb{R}^N \times \mathbb{R}^N$ and $\Lambda_3 \in \mathbb{R}^N$ are denoted by

$$\Lambda_1 = \begin{bmatrix} \xi_{\text{ref}}(\theta_0^*) - \theta_1^*(\xi_{\text{ref}}(\theta_1^*) - \xi_{\text{ref}}(\theta_0^*)) \\ \vdots \\ \xi_{\text{ref}}(\theta_{N-1}^*) - \theta_N^*(\xi_{\text{ref}}(\theta_N^*) - \xi_{\text{ref}}(\theta_{N-1}^*)) \end{bmatrix}, \quad \Gamma_2 = \begin{bmatrix} 1 & 0 & \cdots & 0 \\ 1 & 1 & \cdots & 0 \\ 1 & 1 & \cdots & 1 \end{bmatrix},$$

$$\Lambda_2 = \begin{bmatrix} \xi_{\text{ref}}(\theta_1^*) - \xi_{\text{ref}}(\theta_0^*) & \cdots & 0 \\ & \ddots & \\ 0 & \cdots & \xi_{\text{ref}}(\theta_N^*) - \xi_{\text{ref}}(\theta_{N-1}^*) \end{bmatrix}, \quad \Lambda_3 = \begin{bmatrix} \theta_0 + 1 \\ \vdots \\ \theta_0 + N \end{bmatrix},$$

$$\tag{C.19}$$

and $\Pi_3 = [0_{N \times Nm} \ I_{N \times N}] \in \mathbb{R}^{N \times N(m+1)}$. Moreover, the stacked input reference is defined as $U_{\text{ref}} = \left[u_{\text{ref}}(\theta_0^*)^\top \ \cdots \ u_{\text{ref}}(\theta_{N-1}^*)^\top \right] \in \mathbb{R}^{Nm}$ and we have that

$$H = \begin{bmatrix} H_1 & H_2 \\ H_2^\top & H_3 \end{bmatrix}, \quad F = \begin{bmatrix} F_1 & F_2 \end{bmatrix}, \tag{C.20}$$

with the auxiliary matrices $H_1 \in \mathbb{R}^{Nn \times Nn}$, $H_2 \in \mathbb{R}^N$, $H_3 \in \mathbb{R}^{N \times N}$, $F_1 \in \mathbb{R}^{Nn \times 1}$, and $F_2 \in \mathbb{R}^N$, which are defined as

$$
\begin{aligned}
H_1 &= \Gamma_1^\top \Omega_2^\top \tilde{Q} \Omega_2 \Gamma_1 + \tilde{R}, \\
H_2 &= -\Gamma_1^\top \Omega_2^\top \tilde{Q} \Lambda_2 \Gamma_2, \\
H_3 &= \Gamma_2^\top \Lambda_2^\top \tilde{Q} \Lambda_2 \Gamma_2 + \tilde{T}, \\
F_1 &= 2 S^\top \tilde{Q} \Omega_2 \Gamma_1, \\
F_2 &= -2 S^\top \tilde{Q} \Lambda_2 \Gamma_2,
\end{aligned}
\tag{C.21}
$$

with

$$
\tilde{Q} = \begin{bmatrix} I_{N-1} \otimes Q & 0 \\ 0 & P \end{bmatrix} \in \mathbb{R}^{N(n+1) \times N(n+1)}, \quad \tilde{R} = \begin{bmatrix} I_N \otimes R \end{bmatrix} \in \mathbb{R}^{Nm \times Nm},
\tag{C.22}
$$
$$
\tilde{T} = \begin{bmatrix} I_N \otimes T \end{bmatrix} \in \mathbb{R} \times \mathbb{R},
$$

and $S = \Omega_1 \xi_0 + \Omega_3 E + \Omega_2 U_{-1} - \Lambda_1 - \Lambda_2 \Lambda_3$. The constant $s \in \mathbb{R}$ can be written as

$$
s = S^\top \tilde{Q} S.
\tag{C.23}
$$

Now, we want to define the matrices and vectors appearing in (4.28). More precisely, we have

$$
G = \begin{bmatrix} \tilde{C}_x (\Omega_2 \Gamma_1 \Pi_2 - \Lambda_2 \Gamma_2 \Pi_3) \\ \tilde{C}_u \Gamma_1 \Pi_2 \\ \tilde{C}_{\Delta u} \Pi_2 \end{bmatrix}, \quad r = \begin{bmatrix} \tilde{C}_x S \\ \tilde{C}_u (U_{-1} - U_{\text{ref}}) \\ 0 \end{bmatrix},
\tag{C.24}
$$
$$
d(w) = \begin{bmatrix} \tilde{d}_x - \tilde{C}_x (\Lambda_1 + \Lambda_2 \Lambda_3 + \Lambda_2 \Gamma_2 \Pi_3 w) \\ \tilde{d}_u - \tilde{C}_u U_{\text{ref}} \\ \tilde{d}_{\Delta u} \end{bmatrix}, \quad G_v = \tilde{C}_v \Pi_3 \, d_v = \tilde{d}_v
\tag{C.25}
$$

with $\tilde{C}_x$, $\tilde{C}_u$, $\tilde{C}_v$, $\tilde{d}_x$, $\tilde{d}_u$, and $\tilde{d}_v$ being defined in (C.5) - (C.7) and $\tilde{C}_v$, $\tilde{d}_v$ being defined accordingly.

147

Notation

The following list presents an overview of the most frequently used symbols and acronyms. Precise definitions are given at the symbol's first appearance.

General Notation and Symbols

Sets

$\mathbb{N}$	Set of natural numbers, $0 \in \mathbb{N}$.
$\mathbb{N}_+$	Set of positive natural numbers, $0 \notin \mathbb{N}_+$.
$\mathbb{R}$	Set of reals.
$\mathbb{R}_+$	Set of positive reals $0 \in \mathbb{R}_+$.
$\mathbb{R}_{++}$	Set of strictly positive reals $0 \notin \mathbb{R}_{++}$.
$\mathbb{R}_{++}^n$	Set of n-dimensional vectors with strictly positive real entries.
$\mathbb{S}_+^n$	Set of $n \times n$-dimensional positive semi-definite matrices.
$\mathbb{S}_{++}^n$	Set of $n \times n$-dimensional positive definite matrices.
$[a, b)$	Interval $\mathcal{I} = \{x \in \mathbb{R} : a \leq x < b\}$ for $a, b \in \mathbb{R}$.
$\partial \mathcal{S}$	Boundary for a given set $\mathcal{S}$.
$\mathbb{SO}(n)$	Group of rotations in $\mathbb{R}^n$.

Vectors, matrices, and norms

$I_n, 0_n$	Identity/zero matrix of dimension $n \in \mathbb{N}_+$.
$0_{n \times m}, I_{n \times m}$	$n \times m$ matrix of zeros/ones for $n, m \in \mathbb{N}_+$.
$(\cdot)^\top$	Transposition of a real vector or matrix.
$(\cdot)^i$	ith element or row of a real vector or matrix.
$\lambda_{\min}(A)$	Minimal eigenvalue of a symmetric matrix $A \in \mathbb{S}^n$.
$\lambda_{\max}(A)$	Maximal eigenvalue of a symmetric matrix $A \in \mathbb{S}^n$.
$\|x\|$	Euclidean norm of vector $x \in \mathbb{R}^n$ with $\|x\| = \sqrt{x^\top x}$.
$\|x\|_M$	Weighted Euclidean norm of vector $x \in \mathbb{R}^n$ with $\|x\| = \sqrt{x^\top M x}$.
$\otimes$	Kronecker matrix product.

Functions and derivatives

$f : D \to C$	The function $f(\cdots)$ maps from its domain D into the set C.
C^k	The set of k-times continuously differentiable functions, $k \in \mathbb{N}$.
$\nabla f(\bar{x})$	Gradient of a function $f(x)$, $f : D \to \mathbb{R}$, evaluated at $\bar{x} \in D$.
$\nabla^2 f(\bar{x})$	Hessian of a function $f(x)$, $f : D \to \mathbb{R}$, evaluated at $\bar{x} \in D$.
$\lfloor x \rfloor$	$\lfloor x \rfloor = \max\{i \in \mathbb{Z} \mid i \le x\}$ with $x \in \mathbb{R}$ belonging to the down rounding function.
$\mathcal{K}$	A function $\alpha : \mathbb{R}_+ \to \mathbb{R}_+$ is of class $\mathcal{K}$, if it is $\alpha(0) = 0$, continuous and strictly increasing.
$\mathcal{K}_\infty$	A function $\alpha : \mathbb{R}_+ \to \mathbb{R}_+$ is of class $\mathcal{K}_\infty$, if it is of class $\mathcal{K}$ and $\alpha(r) \to \infty$ as $r \to \infty$.
$\mathcal{KL}$	A function $\beta : \mathbb{R}_+ \times \mathbb{R}_+ \to \mathbb{R}_+$ is of class $\mathcal{KL}$ if for each fixed $s \in \mathbb{R}_+$, it holds that $\beta(r,s)$ belongs to class $\mathcal{K}$. Moreover, for each fixed $r \in \mathbb{R}_+$, the function $\beta(r,s)$ is decreasing with respect to s and it holds that $\lim_{s \to \infty} \beta(r,s) = 0$.

Acronyms

DARE	Discrete-time algebraic Riccati equation.
KKT	Karush-Kuhn-Tucker.
LQR	Linear quadratic regulator.
LTI/LTV	Linear time-invariant/time-varying.
(N)MPC	(Nonlinear) Model predictive control.
rbMPC	Model predictive control based on relaxed barrier functions.
ECU	Electronic control unit.
EBS	Electronic braking system.
EPS	Electronic power steering.
ARX	Auto regressive with exogenous input.
ICR	Instantaneous center of rotation.
LQR	Linear quadratic regulator.
ABS	Anti blocking system.

Bibliography

A. Aguiar and J. Hespanha. (2007). Trajectory-Tracking and Path-Following of Underactuated Autonomous Vehicles with Parametric Modeling Uncertainty. *IEEE Transactions on Automatic Control, 52*(8), 1362-1379.

A. Aguiar, J. Hespanha and P. Kokotovic. (2005). Path-following for nonminimum phase systems removes performance limitations. *IEEE Transactions on Automatic Control, 50*(2), 234-239.

A. Ahmadi and R. Jungers. (2018). SOS-Convex Lyapunov Functions and Stability of Difference Inclusions. *https://arxiv.org/pdf/1803.02070.pdf*.

A. Sabelhaus et al. (2018). Trajectory Tracking Control of a Flexible Spine Robot, With and Without a Reference Input. *https://arxiv.org/abs/1808.08309*.

A. Wills and W. Heath. (2002). A recentered barrier for constrained receding horizon control. *In Proceedings of the 2002 American Control Conference, 5*, 4177-4182.

B. Gutjahr, L. Gröll and M. Werling. (2017). Lateral vehicle trajectory optimization using constrained linear time-varying MPC. *IEEE Transactions on Intelligent Transportation Systems, 18*(6), 1586-1595.

B. Koopman. (1931). Hamiltonian systems and transformation in Hilbert space. *Proceedings of the National Academy of Sciences of the United States of America, 17*, 315–318.

C. Aguilar-Avelar and J. Moreno-Valenzuela. (2017). A MRAC Principle for a Single-Link Electrically Driven Robot with Parameter Uncertainties. *Complexity*.

C. Canudas and R. Roskam. (1991). Path following of a 2-DOF wheeled mobile robot under path and input torque constraints. *In Proceedings of the IEEE International Conference on Robotics and Automation, 2*, 1142 - 1147.

C. Ebenbauer, F. Pfitz and S. Yu. (2021). Control of Unknown (Linear) Systems with Receding Horizon Learning. *Proceedings of the 3rd Conference on Learning for Dynamics and Control, 144*, 872-879.

C. Feller. (2017). Relaxed Barrier Function Based Model Predictive Control [PhD thesis, University of Stuttgart]. *Logos Verlag Berlin*.

C. Feller and C. Ebenbauer. (2016). A stabilizing iteration scheme for model predictive control based on relaxed barrier functions. *Automatica, 80*, 328-339.

C. Feller and C. Ebenbauer. (2017). Relaxed Logarithmic Barrier Function Based Model Predictive Control of Linear Systems. *IEEE Transactions on Automatic Control, 62*(3), 1223-1238.

C. Feller and C. Ebenbauer. (2020). Sparsity-Exploiting Anytime Algorithms for Model Predictive Control: A Relaxed Barrier Approach. *IEEE Transactions on Control Systems Technology, 28*(2), 425–435.

C. Münzing. (2017). *A novel functional exosystem observer with application to relaxed-barrier MPC* [Master thesis, University of Stuttgart, Advisor: C. Ebenbauer and C. Feller].

C. Olsson. (2015). *Model Complexity and Coupling of Longitudinal and Lateral Control in Autonomous Vehicles Using Model Predictive Control* [Master thesis, KTH Royal Institute of Technology].

D. Laila and A. Astolfi. (2004). Input-to-state stability for parameterized discrete-time time-varying nonlinear systems with applications. *In Proceedings of the 5th Asian Control Conference, 1*, 274-282.

D. Limon et al. (2008). MPC for tracking piecewise constant references for constrained linear systems. *Automatica, 44*(9), 2382-2387.

D. Ruscio. (2013). Model Predictive Control with Integral Action: A simple MPC algorithm. *Modeling, Identification and Control, 34*(3), 119-129.

D. Simon, J. Löfberg and T. Glad. (2014). Reference Tracking MPC Using Dynamic Terminal Set Transformation. *IEEE Transactions on Automatic Control, 59*(10), 2790-2795.

E. Gilbert and T. Tan. (1991). Linear systems with state and control constraints: The theory and application of maximal output admissible sets. *IEEE Transactions on Automatic Control, 36*(9), 1008-1020.

E. Yurtserver et al. (2020). A survey of autonomous driving: Common practices and emerging technologies. *IEEE Access, 8*, 58443-58469.

F. Pfitz and M. Schaefer. (2022). Automated Endurance Testing and an Outlook to AI. *In Proceedings of the 12th International Munich Chassis Symposium*, 183-199.

F. Pfitz, C. Ebenbauer and M. Braun. (2019). Relaxed barrier MPC for Reference Tracking: Theoretical and experimental studies. *VDI-Berichte 2349 AUTOREG, Mannheim Germany*, 97-110.

F. Pfitz, X. Hu and C. Ebenbauer. (2021). Relaxed Barrier MPC for Path Following in Constrained Environments. *In Proceedings of the 29th Mediterranean Conference on Control and Automation*, 872-879.

G. Betti, M. Farina and R. Scattolini. (2012). An MPC algorithm for offset-free tracking of constant reference signals. *51st IEEE Conference on Decision and Control*, 5182-5187.

G. Goodwin and K. Sin. (2009). *Adaptive Filtering Prediction and Control*. USA: Dover Publications, Inc.

G. Hoffmann and S. Waslander. (2008). Quadrotor helicopter trajectory tracking control. *AIAA Guidance, Navigation and Control Conference and Exhibit*.

G. Pannocchia and J. Rawlings. (2001). The velocity algorithm LQR: a survey. In *Technical report 2001-01*.

G. Pannocchia, M. Gabiccini, A. Artoni. (2015). Offset-free MPC explained: novelties, subtleties, and applications. *IFAC PapersOnLine, 48*(23), 342-351.

G. Tao. (2014). Multivariable adaptive control: A survey. *Automatica, 50*(11), 2737 - 2764.

H. Bauschke and J. Borwein. (2001). Joint and Separate Convexity of the Bregman Distance. *Studies in Computational Mathematics, Elsevier, 8*, 23–36.

H. Burg and A. Moser. (2017). *Handbuch Verkehrsunfallrekonstruktion: Unfallaufnahme, Fahrdynamik, Simulation*. Springer Vieweg.

H. Edwards and Y. Lin and Y. Wang. (2000). On input-to-state stability for time varying nonlinear systems. *In Proceedings of the 39th IEEE Conference on Decision and Control, 4,* 3501–3506.

H. Pacejka. (2012). *Tire and Vehicle Dynamics.* Butterworth-Heinemann.

I. Adelwahed, A. Mbarek and K. Bouzrara. (2017). Adaptive MPC based on MIMO ARX-Laguerre model. *ISA Transactions, 67,* 330-347.

I. Batkovic et al. (2020). Safe trajectory tracking in uncertain environments. *https://arxiv.org/abs/2001.1160.*

J. Hauser and A. Saccon. (2006). A Barrier Function Method for the Optimization of Trajectory Functionals with Constraints. *In Proceedings of the 45th IEEE Conference on Decision and Control,* 864-869.

J. Hauser and R. Hindmann. (1995). Maneuver regulation from trajectory tracking: Feedback linearizable systems. *3rd IFAC Symposium on Nonlinear Control Systems Design, 28*(14), 595-600.

J. Kong et al. (2015). Kinematic and dynamic vehicle models for autonomous driving control design. *IEEE Intelligent Vehicles Symposium (IV),* 1094-1099.

J. Maxwell. (1868). On Governors. *In Proceedings of the Royal Society in London.*

J. Morales et al. (2009). Pure-pursuit reactive path tracking for nonholonomic mobile robots with a 2D laser scanner. *Eurasip Journal on advances in signal processing,* 1-10.

J. Rawlings, D. Mayne and M. Diehl. (2019). *Model Predictive Control: Theory, Computation and Design.* Nob Hill Publishing.

J. Yang and J. Kim. (1999). Sliding Mode Control for Trajectory Tracking of Nonholonomic Wheeled Mobile Robots. *IEEE Transactions on Robotics and Automation, 15*(3), 578-587.

M. Althoff. (2018). Common Road: Vehicle Models. *Technische Univeristät München, 85748 Garching, Germany.*

M. Bakosova and J. Oravec. (2014). Robust MPC of an unstable chemical reactor using the nominal system optimization. *Acta Chimica Slovaca, 7*(2), 87-93.

M. Bargende et al. (2019). 19. Internationales Stuttgarter Symposium. *Springer Vieweg.*

M. Benosman. (2016). *Learning-Based Adaptive Control: An Extremum Seeking Approach – Theory and Applications.* Elsevier Science.

M. Forbes et al. (2015). Model Predicitve Control in Industry: Challenges and Opportunities. *5th IFAC Conference on Nonlinear Model Predictive Control, 48*(8), 531-538.

M. Gharbi, B. Gharesifard and C. Ebenbauer. (2020). Anytime proximity moving horizon estimation: Stability and regret. *https://arxiv.org/abs/2006.14303.*

M. Korda and I. Mezic. (2017). Linear predictors for nonlinear dynamical systems: Koopman operator meets model predictive control. *https://arxiv.org/abs/1611.03537.*

M. Mitschke and H. Wallentowitz. (1972). *Dynamik der Kraftfahrzeuge.* Springer.

M. Samuel, M. Hussein and M. Binti. (2016). A Review of Some Pure-Pursuit based Path Tracking Techniques for Control of Autonomous Vehicle. *International Journal of Computer Applications, 135,* 35-38.

M. Stephens and M. Good. (2013). Model Predictive Control for Reference Tracking on an Industrial Machine Tool Servo Drive. *IEEE Transactions on Industrial Informatics, 9*(2), 808-816.

N. Saraf and A. Bemporad. (2017). Fast model predictive control based on linear input/output models and bounded-variable least squares. *IEEE 56th Annual Conference on Decision and Control (CDC)*, 1919-1924.

P. Bouffard, A. Aswani and C. Tomlin. (2012). Learning-based model predictive control on a quadrotor: Onboard implementation and experimental results. *IEEE International Conference on Robotics and Automation*, 279-284.

P. Chiaccio. (1990). Exploiting redundancy in minimum-time path following robot control. *American Control Conference*, 2313-2318.

P. Falcone et al. (2007). Predictive Active Steering Control for Autonomous Vehicle Systems. *IEEE Transactions on control systems technology*, 15(3), 566-588.

P. Falcone et al. (2008). Linear time-varying model predictive control and its application to active steering systems: Stability analysis and experimental validation. *International Journal of Robust and Nonlinear Control*, 18(8), 862-875.

P. Tabuada and L. Fraile. (2020). Data-driven Stabilization of SISO Feedback Linearizable Systems. *https://arxiv.org/abs/2003.14240*.

Porsche. (2020). *Proofing ground Weissach*. Retrieved 16.09.2020, from https://www.luftbildsuche.de/info/luftbilder/teststrecke-uebungsplatz-fahrsicherheitszentrum-weissach-baden-wuerttemberg-deutschland-388714.html

Protter, M. H., & Morrey, C. B. J. (2012). *A First Course in Real Analysis*. Springer New York.

R. Attia, R. Orjuela and M. Basset. (2014). Combined longitudinal and lateral control for automated vehicle guidance. *Vehicle System Dynamics*, 52, 261-279.

R. Coulter. (1992). Implementation of the pure pursuit path tracking algorithm. *Carnegie-Mellon University Pittsburgh PA Robotics Institute, CMU-RI-TR-92-01*.

R. Jazar. (2017). *Vehicle Dynamics*. Springer-Verlag New York.

R. Rajamani. (2012). *Vehicle Dynamics and Control*. Springer.

S. Bittanti and M. Campi. (2006). Adaptive Control of Linear Time Invariant Systems: The "Bet on the Best" Principle. *Communications in Information and Systems*, 6(4), 299–320.

S. Boyd and L. Vandenberghe. (2004). *Convex Optimization*. Cambridge University Press.

S. Brunton et al. (2016). Koopman invariant subspaces and finite linear representations of nonlinear dynamical systems for control. *PloS ONE*, 11(2).

T. Carleman. (1932). Application de la theorie des equations integrales lineaires aux systems d'equations differentielles non lineaires. *Acta Mathematica*, 59, 63–87.

T. Faulwasser. (2012). Optimization-based Solutions to Constrained Trajectory-tracking and Path-following Problems [PhD thesis, Otto-von Guericke-Universität Magdeburg].

T. Faulwasser and R. Findeisen. (2011). A model predictive control approach to trajectory tracking problems via time-varying level sets of Lyapunov functions. *In Proceedings of the 50th IEEE Conference of Decision and Control and European Control Conference*, 3381-3386.

T. Fossen et al. (2003). Line-of-sight path following of underactuated marine craft. *6th IFAC Conference on Manoeuvring and Control of Marine Craft*, 36(21), 211-216.

T. Nguyen et al. (2020). Output-Feedback RLS-Based Model Predictive Control. *2020 American Control Conference (ACC)*, 2395-2400.

T. Schanz. (2018). *Relaxed Barrier Function Based Model Predictive Tracking Control of a Sports Car* [Master thesis, Institute for Systems Theory and Automatic Control, University of Stuttgart, Germany, Advisor: F. Pfitz].

U. Maeder and M. Morari. (2010). Offset-free reference tracking with model predictive control. *Automatica, 46*(10), 1469-1476.

U. Maeder, F. Borrelli and M. Morari. (2009). Linear offset-free Model Predictive Control. *Automatica, 45*(10), 2214-2222.

V. Adetola, D. DeHaan and M. Guay. (2009). Adaptive model predictive control for constrained nonlinear systems. *Systems and Control Letters, 58*(5), 320 - 326.

Vehico GmbH. (2020). *ISO Double Lane Change Test.* Retrieved 13.09.2020, from https://vehico.com/index.php/en/applications/iso-lane-change-test

W. Haddad and V. Chellaboina. (2011). *Nonlinear Dynamical Systems and Control: A Lyapunov-Based Approach.* Princeton University Press.

Y. Censor and S. Zenios. (1991). Proximal minimization algorithm with D-functions. *Journal of Optimization Theory and Applications, 73*, 451 - 464.

Z. Jiang and Y. Wang. (2002). A converse Lyapunov theorem for discrete-time systems with disturbances. *Systems and Control Letters, 45*(1), 49-58.